Chemistry
and the
Food System

Chemistry
and the
Food System

A Study by
The Committee on Chemistry and Public Affairs

AMERICAN CHEMICAL SOCIETY
WASHINGTON, D.C.
1980

Library of Congress CIP Data

American Chemical Society. Committee on Chemistry
    and Public Affairs.
    Chemistry and the food system.

    Includes bibliographies and index.
    1. Food industry and trade. 2. Chemical engineering. I. Title.
TP370.5.A43   1980   664   80-11194
ISBN 0-8412-0557-4
ISBN 0-8412-0563-9 pbk.

Copyright © 1980

American Chemical Society

Printed in the United States of America

Contents

Preface

A study of the food system is especially appropriate at this time for a number of reasons, including the growing national and international commitment to improve nutrition, the increasing public interest in the environment and quality of life, and the pressures on food production stemming from population growth, energy shortages, and other socioeconomic factors.

This report has been prepared by the Committee on Chemistry and Public Affairs as part of a program initiated in 1966 to develop reports on public problems involving chemistry. The intent of the program is to contribute to the understanding and solutions of such problems and thereby benefit both the public and the chemical profession. The reports are designed to provide public policy makers with information and recommendations that will lead to sound decisions on key issues. Previous reports have been issued under the titles of *Cleaning Our Environment* (two editions), *Chemistry in the Economy*, and *Chemistry in Medicine*.

The purpose of the present study is to enhance public understanding of the role of chemistry in the food system, a role that reaches from field and farm to table. Such understanding should include a general knowledge of how and why chemical substances are used, as well as what the attendant benefits, disadvantages, problems, and risks may be. It is also important to have an appreciation of how risks can be estimated, monitored, and regulated.

We have tried to take into account both the wants of people and their needs, including a diet that provides good nutrition and is protected from introduction of components that could be harmful. The report describes how a range of chemical substances—including fertilizers, pesticides, plant growth regulators, drugs to control animal diseases, and nutritional supplements—have clearly provided benefits in production, storage, and processing of foods from plants and animals. Some of the newer areas such as fabricated foods, special dietary foods, and unconventional sources of food and feed are discussed.

For convenience, a summary and recommendations are provided in Chapter 1; each subject mentioned in that chapter is developed in more detail in a subsequent chapter. Each chapter considers the potential for further benefits through application of chemistry, including making the best use of various resources. Priorities for research are also indicated. The possibilities for increasing efficiency of plant production through manipulation of such metabolic processes as nitrogen fixation and carbohydrate syntheses are particularly intriguing. In the current conditions of diminishing land, water, and other resources required for expansion of agricultural production, improvements in the efficiency of plant and animal growth must receive more attention.

We have also dealt with recent episodes in which chemicals may have done harm to consumers or the environment, and the lessons learned from them. Protection of the consumer is also taken up in Chapter 5, "Assuring the Wise Use of Chemicals," in which we recommend that benefits as well as risks should be considered in the regulation of the use of chemicals in the food system. We recognize that more work may have to be done in many instances to develop an adequate information base and methodology so that this approach can be applied rationally. Chemistry and biochemistry are valuable tools that enhance our basic understanding of the workings of the food system, and also help us gain better understanding of how various substances are handled in the body.

Although this report is concerned mainly with the situation in the United States and other industrialized countries, there is in fact a global interdependence in these matters. The report makes some references to international aspects, including special considerations for the developing countries where different conditions may call for somewhat different procedures, although the basic principles remain the same. We urge continued international cooperation at the scientific, professional, and governmental levels.

Lester J. Teply
Study Chairman

Robert A. Alberty
Chairman, Committee on
Chemistry and Public Affairs

Contributors

J.C. Bauernfeind (Retired)
Nutrition Research Coordinator
Hoffman-LaRoche, Inc.
Nutley, New Jersey

H.R. Bolin
Western Utilization Research
    Center
Agricultural Research Service
U.S. Department of Agriculture
Berkeley, California

Charles H. Davis
Director, Chemical Development
    Division
National Fertilizer Development
    Center
Muscle Shoals, Alabama

Owen Fennema
Professor of Food Chemistry and
    Chairman, Department of Food
    Science
University of Wisconsin
Madison, Wisconsin

Richard L. Hall
Vice President, Science &
    Technology
McCormick & Co., Inc.
Hunt Valley, Maryland

John E. Halver
U.S. Fish and Wildlife Service
College of Fisheries
University of Washington
Seattle, Washington

Ralph W.F. Hardy
Associate Director
Central Research & Development
    Department
Experimental Station
E.I. du Pont de Nemours & Co.
Wilmington, Delaware

Randolph Hatch
Department of Chemical
    Engineering
University of Maryland
College Park, Maryland

Marcus Karel
Deputy Head, Food Science and
Nutrition Department
Massachusetts Institute of
Technology
Cambridge, Massachusetts

Jack Plimmer
Chief, Organic Chemical
Synthesis Laboratory
Agricultural Environmental
Quality Institute
U.S. Department of Agriculture
Agricultural Research Center
Beltsville, Maryland

D.K. Salunkhe
Department of Nutrition and Food
Sciences
Utah State University
Logan, Utah

Acknowledgments

Advisors

We would like to thank the following for acting as key advisors to the principal contributors:

N.J. Benevenga
Professor of Nutrition &
  Biochemistry
University of Wisconsin
Madison, Wisconsin

D.G. Crosby
University of California
Davis, California

John W. Frankenfeld
Exxon Research & Engineering
  Co.
Linden, New Jersey

G.K. Kohn
Zoecon Corp.
Palo Alto, California

M.L. Leng
Dow Chemical, U.S.A.
Midland, Michigan

Donald L. McCune
Managing Director
International Fertilizer
  Development Center
Florence, Alabama

Walter Mertz
Chairman, Nutrition Institute
U.S. Department of Agriculture
Beltsville, Maryland

Sanford A. Miller
Professor, Department of Nutrition
  & Food Science
Massachusetts Institute of
  Technology
Cambridge, Massachusetts
(Presently Director, Bureau of
  Foods
Food & Drug Administration
Washington, D.C.)

J.D. Nickerson
USS Agri-Chemicals
Atlanta, Georgia

Kenneth L. Parks
Agrico-Chemical Co.
Bartow, Florida

Willard B. Robinson
Head, Institute of Food Science
New York State College of
   Agriculture & Life Science
Cornell University
Geneva, New York

Herbert P. Sarett
Vice President, Nutritional
   Science Resources
Mead Johnson Research Center
Evansville, Indiana

W.M. Upholt
Environmental Protection Agency
Washington, D.C.

Alvin Weiss
Department of Chemical
   Engineering
Worcester Polytechnic Institute
Worcester, Massachusetts

Thomas L. Welsh
Vice President, Research &
   Development
Miles Laboratories, Inc.
Grocery Products Division
Chicago, Illinois

M.T. Wu
Department of Nutrition & Food
   Sciences
Utah State University
Logan, Utah

Outside Reviewers

We would like to acknowledge the helpful comments and suggestions made by the following in the course of their reviews of all or parts of the manuscript:

Dr. James Austin
Harvard Institute of International
   Development
Harvard University
Cambridge, Massachusetts

Dr. Orville G. Bentley
Dean, College of Agriculture
University of Illinois
Urbana, Illinois

Dr. Darshan S. Bhatia
Director, Corporate R & D
Coca Cola Co.
Atlanta, Georgia

Dr. Douglas V. Frost
Consultant in Nutrition
   Biochemistry
Schenectady, New York

Dr. Dee M. Graham
Del Monte Research Center
Walnut Creek, California

Dr. Hartley W. Howard (Retired)
Borden Co.
New Milford, Connecticut

Dr. James Johnston
Deputy Director for Agricultural
  Sciences
Rockefeller Foundation
New York, New York

Dr. Carl Krieger (Retired)
Campbell Soup Co.
Wynnewood, Pennsylvania

Dr. Paul LaChance
Professor of Nutritional
  Physiology
Department of Food Science
Rutgers–The State University
New Brunswick, New Jersey

Dr. Stanley Lebergott
Institute for Advanced Studies
Princeton University
Princeton, New Jersey

Dr. Gilbert A. Leveille
Michigan State University
East Lansing, Michigan

Dr. Louis G. Nickell
Velsicol Chemical Corporation
Chicago, Illinois

Dr. Bernard L. Oser
Consultant in Food Regulations
New York, New York

Dr. Gary L. Rumsey
U.S. Department of the Interior
Fish and Wildlife Service
Cortland, New York

Dr. Howard Schneider
Emeritus Professor of
  Biochemistry & Nutrition
University of North Carolina
  School of Medicine
Chapel Hill, North Carolina

Dr. Bernard Schweigert
Chairman, Department of Food
  Science & Technology
University of California
Davis, California

Dr. Frank M. Strong (Retired)
Biochemistry Department
University of Wisconsin
Madison, Wisconsin

Dr. Daniel Wang
Department of Nutrition & Food
  Science
Massachusetts Institute of
  Technology
Cambridge, Massachusetts

Dr. Israel Zelitch
Head, Biochemistry
Connecticut Agricultural
  Experimental Station
New Haven, Connecticut

Committee on Chemistry and Public Affairs (1979)

Robert A. Alberty, Chairman
Dean of Science
Massachusetts Institute of
  Technology
Cambridge, Massachusetts

William J. Bailey
Research Professor of Chemistry
University of Maryland
College Park, Maryland

Herman L. Finkbeiner
Manager, Planning & Resources
Materials Science Engineering
General Electric Co.
Schenectady, New York

Anna J. Harrison
Professor of Chemistry
Mount Holyoke College
South Hadley, Massachusetts

W. Lincoln Hawkins
Research Director
Plastics Institute of America, Inc.
Hoboken, New Jersey

Thomas J. Kucera
Vice President & Technical
  Director
APECO Corporation
Elk Grove Village, Illinois

Blaine C. McKusick
Assistant Director
Haskell Laboratory for Toxicology
  and Industrial Medicine
E.I. du Pont de Nemours & Co.,
  Inc.
Wilmington, Delaware

E. Gerald Meyer
Vice President for Research
University of Wyoming
Laramie, Wyoming

Alan C. Nixon
Consultant
Berkeley, California

Don I. Phillips
Special Science Advisor
North Carolina Board of Science &
  Technology
Office of the Governor
Raleigh, North Carolina

Glenn T. Seaborg
University Professor of Chemistry
Lawrence Berkeley Laboratory
University of California
Berkeley, California

William P. Slichter
Executive Director
Research, Materials Science &
    Engineering Division
Bell Laboratories
Murray Hill, New Jersey

Gardner W. Stacy
Department of Chemistry
Washington State University
Pullman, Washington

Philip W. West
Professor of Chemistry
Department of Chemistry
Louisiana State University
Baton Rouge, Louisiana

Linda S. Wilson
Associate Vice Chancellor for
    Research
University of Illinois
Urbana, Illinois

M. Kent Wilson
Director, Office of Planning &
    Resources Management
National Science Foundation
Washington, D.C.

David C. Young
Process Development
Dow Chemical Co.
Midland, Michigan

Consultants

William O. Baker
President
Bell Laboratories
Murray Hill, New Jersey

Herbert E. Carter
Head of Biochemistry
University of Arizona
Tucson, Arizona

Charles G. Overberger
Vice President for Research
University of Michigan
Ann Arbor, Michigan

Staff Liaison

Annette T. Rosenblum
Assistant to the Director
Department of Public Affairs
American Chemical Society
Washington, D.C.

Summary and Recommendations

Of the world's total population of over 4 billion, at least 450 million suffer from hunger and malnutrition. Most of these people live in the developing countries. But even in the developed nations such as the United States, substantial numbers of people have dietary problems. Some eat too much or have diets of poor quality, or both, while others do not have enough to eat.

By the end of this century, when the world's population will probably be 6 billion or more, an even larger number of people will be inadequately nourished unless more food is produced. But merely producing more food is not enough. Also needed are appropriate distribution of food and an environment conducive to good nutrition and health, including adequate housing, sanitation, clean air and water, and control of infectious diseases. More effort must also be made to minimize food losses during storage and distribution.

Americans expect long, healthy lives and abundant supplies of wholesome and safe food. But many, more consciously and explicitly, also want their consumption of food to contribute as much as possible to the quality of their lives—physically, mentally, and socially. The aesthetic aspects of eating should never be forgotten, for although food is essential to physical well-being, it is also a significant aspect of family life and culture. Alternatives to fresh foods (and other traditional foods as well) should, as much as possible, not detract from the pleasure of eating. Most Americans not only want to consume enough nutrients and to enjoy eating, but also want to control their intake of calories, fats, salt, and carbohydrates in the interest of lowering the risk of such health problems as obesity, hypertension, cardiovascular disease, and diabetes.

People of developing countries also have rising aspirations. Their most pressing problem is to provide at least minimal food intake and nutrition for the general population and to guard against future food shortages. This problem was highlighted at the World Food Conference of 1974 in Rome where, after many years of neglect, governments joined together to move toward giving the problems of inadequate food supplies and poor nutrition the international attention they deserve. The World Food Council is promoting implementation of the Conference's recommendations, which call for greater international cooperation in improving food production and distribution.

Increasing population and greater world interdependence are major factors that affect the world's food supplies. Although each country will still continue to make its own food policy, there is need for some international standardization as, for example, in the establishment of quality standards for foods in international trade under the Codex Alimentarius of the Food and Agriculture Organization and the World Health Organization.

The solution of food and nutrition problems is heavily dependent on research in many scientific disciplines. Among the most important of these are chemistry and biochemistry, which touch every phase of the food system from field to table and thus can point the way to practical improvements throughout the system. Chemistry can be applied to increase food supplies and to use them more efficiently. It also can enhance our understanding of how foods, which are composed of chemicals, are handled in the body and provide the analytical tools necessary for progress in many areas of food and nutrition research. Chemistry, however, cannot do the job alone. For optimum results, chemical research must be integrated with research in other disciplines such as biology, genetics, agronomy, and economics.

Together, the food industry and the chemical industry have brought the results of research (carried out by university and government scientists as well as industrial scientists) to the consumer in a wide array of nutritious, safe, economical, and convenient foods. Vitamins, minerals, and amino acids increase the nutritional value of certain foods; flavoring compounds improve taste and aroma; antioxidants and preservatives inhibit spoiling of food or deterioration of flavor; emulsifiers, stabilizers, and thickeners modify texture; leavening and maturing agents improve baked goods; sequestrants, humectants, and anticaking and firming agents maintain certain desirable characteristics of foods; and a wide selection of packaging materials protects foods while they are being shipped and distributed.

The contributions of science and technology are limited by social, political, and economic constraints. Still, science and technology have contributed significantly in the past to increasing supplies of nutritious and wholesome food and should be able to continue to do so. However, the benefits resulting from applying science and technology to the food system must always be weighed against possible disadvantages and risks.

This report discusses the chemical aspects of the production and processing of food. The remainder of this chapter summarizes the major subjects covered in the report and makes recommendations that, if implemented, should help improve the food system. In some cases, the recommendations call for substantial new funding. Funding agencies should find these recommendations useful in identifying those areas where increased efforts are likely to be most productive.

FERTILIZERS

Fertilizer use in the United States has expanded fourfold since 1950. Today, as a result of application of chemical nutrients (primarily compounds containing nitrogen, phosphorus, and potassium) to the soil, U.S. food production is 30 to 40% higher than it would otherwise be. To be most effective, fertilizers must be part of a carefully planned system for managing all controllable factors involved in productivity.

Practically all nitrogen fertilizer is derived from ammonia, which in turn is produced from natural gas and nitrogen from the atmosphere. If new plants come into operation on schedule, supplies of nitrogen fertilizers should be adequate until the early 1980s. Dwindling natural gas resources, however, are expected to increase costs. The most promising long-term alternative appears to be the synthesis of ammonia from coal.

Almost all phosphorus fertilizer is derived from phosphate rock. About three-fourths of U.S. phosphate fertilizer comes from rock mined and processed in central Florida. Resources appear adequate for hundreds of years, although it will be necessary in the future to mine resources of lower quality.

About 80% of the potassium used in the United States is derived from mines in Saskatchewan, and at present rates of consumption, no shortages are likely to develop.

In addition to dwindling reserves of natural gas and high-quality phosphate minerals, a number of other chemical fertilizer problems can be identified:

- *Fertilizer losses in the field.* Only about half of the fertilizer applied to the field is actually used by crops. Much of both nitrogen and potash fertilizers is lost to the crop as they dissolve in water and move through the soil from the point of application. Also, soil bacteria convert some nitrogen fertilizer into gases that can escape to the atmosphere.

- *Energy requirements.* About 70% of the energy needed to produce fertilizers is consumed in making ammonia. Significant amounts of energy are also used in production of phosphoric acid; transportation, storage, and handling of raw materials, intermediates, and products; and drying of granular fertilizers (which account for 70% of U.S. production). Ways must be found to decrease energy usage, which would also decrease pollution.
- *Environmental concerns.* The chemical fertilizer industry has made substantial improvements in its operations in recent years to meet pollution regulations. Probably the most important opportunities for further improvement are related to finding more effective ways to recycle effluents so that they are not discharged into the environment and to decreasing energy use.

Recommendations

We recommend that:

1. Alternative sources of ammonia be examined to ensure adequate supplies.
2. Technology be developed to make more efficient use of fertilizers. Among the possibilities are nitrogen fertilizers that control nitrification or that make plant nutrients available to plants on schedule with needs.

PEST CONTROL

Pesticides—including herbicides, insecticides, fungicides, nematicides, miticides, acaricides, viricides, bactericides, fumigants, and pest attractants and repellents—have increased agricultural productivity throughout the world. Since World War II, a variety of synthetic organic pesticides, among them organic chlorine compounds, organophosphates, and pyrethroids, have been produced in large volumes. Some of their unintentional effects are causing increased concern. Hence, federal laws have been tightened so that all aspects of pesticide use, including licensing, application, storage, and disposal, are regulated.

Production of herbicides, the largest class of pesticides, has increased remarkably in the United States in the past few decades. In 1976, American farmers applied herbicides to 200 million acres of land, up from about

53 million acres in 1959. Demand for herbicides is likely to increase further. In addition, new materials will be needed to replace chemicals found to present unacceptable risks to man and the environment.

Production of insecticides, another large group of pesticides, also is increasing rapidly. The emergence of insect species resistant to existing pesticides is a major concern, which may be solved, in part, by greater use of "integrated pest management." In this approach, the natural forces that regulate pest populations are combined with improved farming methods, use of resistant crop varieties, biological controls, and application of chemicals. With integrated pest management, pesticides will continue to be used, but in smaller amounts.

Fumigants are growing in importance. Much of the world's food is stored for long periods and must be protected from the depredation of pests. The damage is especially great in the tropics and subtropics.

Synthesizing a new pesticide and proving its safety so it can be registered by the Environmental Protection Agency can take 15 to 20 years and an investment of $10–15 million. As a result, only products with a large potential market can be developed economically. It is not economically feasible to develop pesticides for use against unusual pests on major crops or against all important pests on economically less significant crops.

Japan, the United States, and the countries of Western Europe use far more pesticides than the rest of the world combined. Demand is increasing worldwide, however, and is being met in part by exports from the United States and Japan. Production is also increasing in Asia and Latin America, usually with technology developed in the United States and Western Europe. Transfer of such technology can pose a number of problems. Many pesticides that are easily and cheaply manufactured can become environmental pollutants or create disposal problems. Moreover, the majority of U.S. pesticides were developed for use in temperate or subtropical zones. Hence, use patterns must be developed for tropical areas, where there is a particularly pressing need for the developing nations to increase food production. Increasing trade in pesticides calls for greater international cooperation in registration requirements, specifications, tolerance levels, and analytical methodology.

Recommendations

We recommend that:

1. Studies be continued to increase the understanding of the chemical and biochemical components of integrated pest management to permit its wider use in the U.S. food production system.

2. Financial and regulatory incentives be provided to encourage devel-
    opment of pesticides to protect major crops against unusual pests and
    economically less significant crops against all important pests.
3. Improved fumigants or other chemicals be developed to reduce food
    losses during storage.
4. U.S. government, industry, and academic scientists cooperate in
    helping developing nations use pesticides safely and effectively.
5. International cooperation be increased in such matters as registering
    pesticides, establishing specifications, setting tolerance levels, and
    developing uniform analytical methodology.

CROP PRODUCTION

All of the world's food supply is derived from plants (some indirectly
after being fed to animals that can convert plants to meat, eggs, and dairy
products).   Quantitatively, the most important of these plants are cereal
grains such as rice, wheat, and corn, and grain legumes (including soy and
other beans, peas, and lentils).   To meet the world's future food needs,
production of cereal grains must double by the beginning of the 21st century,
while production of grain legumes must quadruple.

Crop production can be increased either by cultivating more land or by
improving yields.   Suitable land is becoming less plentiful, making improved yields, especially in developing countries, essential.   Using all
resources more efficiently will help improve yields.   But, in addition, there
is great potential in new technology that can modify the basic physiological
and biochemical processes by which plants grow.   The natural evolutionary
process results in enhanced ability to survive, not necessarily more production of food for man.   Hence, changing basic plant processes has the potential to increase food supplies.

Breeding by conventional techniques is a slow process, and crosses
between one genera and another (such as wheat with soybean) are precluded.   In the future, plant genetics is likely to depend heavily on the
molecular approach, using tools such as mutation breeding, cell culture,
fusion of vegetative cells, and recombinant DNA.   These tools may be able
to speed the development of improved crop varieties.   They may also help
increase the genetic diversity of plants, which provides a pool on which
plant breeders can draw to develop new varieties with certain desired characteristics.

Photosynthesis is responsible for producing up to 95% of a plant's dry weight.   Photosynthesis is a promising research target because most food crops convert less than 1% of the sunlight reaching the plant's leaves into food.   Crop production is determined by the gross uptake of carbon dioxide during photosynthesis less the carbon dioxide lost during respiration, which occurs in all green plants and algae.   There is evidence that not all of this respiration is essential to the plant, so reducing respiration might improve yields.

In the past quarter century, scientists have discovered a second and apparently wasteful form of plant respiration known as photorespiration (because it occurs at a high rate in most plants when they are exposed to light).   If such plants could be bred to have a lower rate of photorespiration, and hence a higher rate of photosynthesis, yields might be increased.

Nitrogen occurs in every protein molecule and many other molecules that form living tissues and therefore must be supplied to plants.   However, most plants can only assimilate nitrogen when it is combined ("fixed") with other elements.   The nitrogen found in soil can be fixed by certain bacteria. To supplement soil nitrogen, fertilizer is applied to crops.   Adequate food supplies for the world's growing population will require large amounts of nitrogen fertilizer.   The demand could be met by building new plants using present technology.   However, the problems of such expansion, including the cost and availability of fossil fuels, have focused attention on other alternatives.   One alternative is to develop new catalysts that would permit production of ammonia from atmospheric nitrogen at lower temperatures and pressures, thereby reducing cost and energy consumption.   Research on this is still at an early stage and funding is modest.   Hence, large-scale development is probably decades away.

Another way of meeting nitrogen requirements is to exploit nitrogen-fixing bacteria and blue-green algae that live in fields, forests, and oceans. Worldwide, 50% more nitrogen is fixed in agricultural soils than is added as fertilizer; a major part is fixed by bacteria that establish a symbiotic association with legumes.   Since biblical times, legumes have been plowed back in the soil to increase its nitrogen content.   However, the practice is generally not economically competitive in industrialized countries.   Studies are underway to enhance nitrogen fixation in legumes and extend it to such cereal grains as corn and wheat.   Other research aims at transferring the genetic information for fixation directly to plants, thus eliminating the complexities of melding two biological systems—bacteria and the host plant.In both biological and nonbiological nitrogen fixation, there are many promising areas of fundamental research that with adequate funding might lead to practical technologies.

Chemical and biological researchers now hope to increase crop harvests by using plant growth regulators, organic chemicals that beneficially alter the plant's life processes or structures. Initially, regulation of plant growth was considered the exclusive domain of certain naturally occurring plant hormones. Now scientists recognize that any factor that limits yields is subject to manipulation, encouraging the prospects that in the future, plant growth regulators may make as big a contribution to increased crop production as fertilizers, pesticides, and improved plant varieties have in the past.

Recommendation

We recommend that:

Fundamental research be expanded on the basic physiological and biochemical processes of crop production, especially photosynthesis and nitrogen fixation.

ANIMAL PRODUCTION

Animals are the direct source of more than half of the total nutrients in the average U.S. diet. Globally, they provide about a quarter of the average diet. It appears that at least to the year 2000, the United States can produce enough animal products to meet domestic needs and still have surpluses for export. Animals are fed mainly on cereal grains, soybeans, and other edible feeds; on forages (grass, hay, crop residues); and on other materials such as by-products and wastes not consumed by humans. If more edible feeds have to be used for human consumption, animals can be fed larger amounts of nonedible materials.

The past 50 years have seen great progress in animal nutrition. Many animals that once were fed almost entirely on feeds available in fields or barnyards are now fed at least in part on scientifically blended rations containing vitamins, minerals, and amino acids. These rations, designed to promote nutrition, health, and efficient production, are widely used in the poultry industry.

Alone among animals, ruminants (cattle, sheep, and goats) can convert simple nitrogen compounds (via microorganisms in the rumen) into much of the protein they require in their diet, thus reducing the need for edible materials containing nitrogen. The most widely used source of nonprotein nitrogen is urea, but urea is poorly utilized by animals fed largely on forage.

Chemists can make a major contribution to animal production by developing new sources of nonprotein nitrogen that are effective, safe, and inexpensive to use in such diets.

Since the early 1950s, a number of drugs have been added to animal feeds to control disease or stimulate growth.   Among those nonnutritive additives are hormones, antibiotics, sulfonamides, nitrofurans, and arsenicals.   Many have been of considerable value in increasing meat production.   However, in recent years, the use of diethylstilbestrol (DES), a hormone, has been questioned because any residue in edible tissues might cause human cancer.   The use of penicillin and the tetracyclines has been questioned because they may create resistance in bacteria to medications often used to treat human illnesses.   Therefore, the Food and Drug Administration (FDA) has banned the use of DES and has proposed restricting the use of penicillin and the tetracyclines in animal feeds.

Recommendations

We recomend that:

1. Feed resources be increased through processing of nonedible materials to enchance their value as animal feeds.
2. Better sources of nonprotein nitrogen be developed for use in ruminant diets.
3. Efforts be continued to find acceptable growth stimulants for use in animal feeds.

FISH PRODUCTION

Aquaculture—the cultivation and harvest of both freshwater and marine aquatic species—has developed more slowly than have other sources of food in the United States.   In 1975, the United States accounted for only slightly over 1% of total world production of fish.   Developing nations in the tropics and subtropics have great potential for aquaculture because of favorable water temperatures and low labor costs.   However, little research has been done on controlled breeding and rearing of most of the brackish water species found in those areas.

Fish and crustacea reared by modern production methods are fed diets that contain most of the traditional nutrients as well as ingredients peculiar to the needs of aquaculture, such as binders, antioxidants, growth promoters, and attractants.   The problems posed in raising fish are unique because fish are confined to water.   First, feeds must perform well in water.   Second, fish are especially sensitive to pesticides and heavy metal ions.   However, fish can readily metabolize the nucleic acids in single-cell protein (SCP), while humans can metabolize only limited amounts.   Fish are also efficient converters of feed into proteins.   For example, trout produce almost 20 times as much protein from the calories they consume as do beef cattle.

There are a number of opportunities for chemistry to contribute to the growth of aquaculture.   There are as yet no satisfactory larval feeds because present technology cannot produce particles small enough.   Better antioxidants are also required to extend the life of feeds and attractants to increase their acceptability.   Further, disease control must be improved through use of drugs.   Also, sex-controlling chemicals may be useful to (1) improve fish quality through delaying maturation in certain cases and (2) promote more efficient use of water biospheres in which several species of fish live together.

Recommendations

We recommend that:

1. Research be expanded to improve fish feeds, control fish diseases, and develop sex-controlling chemicals.
2. New technology be developed to exploit the potential of aquaculture, especially in the tropic and subtropic regions of the world.

ADDITIVES IN FOOD

Additives are used in foods for several reasons:

● To facilitate processing, handling, distribution, and preparation in the home.
● To control chemical, physical, and microbiological changes.
● To extend shelf life.
● To improve sensory and nutritive properties.

Many additives come from natural sources.   For example, the lecithin in bread sold in supermarkets (added to improve volume, uniformity, and fineness of grain) is extracted from soybeans and corn.   Overall, foods or food components from natural sources constitute well over 99% by weight of the American diet.

The use of additives in the U.S. food industry has grown hand in hand with a number of national developments, including population growth, urbanization, increasing labor costs, greater concern for public health, the consumer's desire for more varied foods, and the need for special dietary foods.   The use of additives has had a great impact on the variety and cost of foods available to the American consumer.   Without additives, there would be few so-called convenience foods, and some fresh fruits and vegetables would not be available year-round.   Other foods would cost more— for example, bread made without preservatives has a short shelf life and so is more expensive to distribute and sell than bread with preservatives. However, additives can have undesirable effects—for instance, color, flavor, and texture can be changed.   In addition, some people fear that additives are used excessively and cause long-term health problems.

FDA regulates the use of additives in food under the Food, Drug, and Cosmetic Act and its 1954–1960 Amendments.   The 1958 amendment includes the Delaney Clause, which bans use of any material found through tests to cause cancer in man or animal.   FDA requires that the manufacturer or industrial user of an additive provide evidence that it is safe in its intended uses.

HANDLING AND STORING OF FRESH FOODS

At some stage in their growth, foods reach an optimum quality for human consumption.   After that, certain chemical processes continue in the food. Eventually, microorganisms cause fresh foods to decompose, unless their growth is prevented.

The food industry applies plant growth regulators to fruits and vegetables during and immediately after harvest.   These chemicals serve a variety of purposes, including extending marketable life and regulating ripening.   Among widely used plant growth regulators are gibberellic acid, which delays ripening of certain fruits and vegetables, and ethephon, which hastens ripening.

Most losses that occur when fresh fruits and vegetables are stored result from the action of fungi and bacteria, but undesirable changes due to enzymatic or other chemical reactions can also occur—for example, vegetables can lose their color, shrivel, sprout, or lose leaves.   In the United States, about 40% of the produce harvested spoils before it is consumed. The most serious infestations can quickly spoil an entire crate.   Sulfur dioxide gas, which kills microorganisms even at low concentrations, is widely used on fruits to prevent such spoilage. This treatment, in combination with reduced temperatures, maintains the appearance, flavor, and texture of fresh fruits.

Recommendation

We recommend that:

Studies can be expanded to develop improved methods to protect fresh fruits and vegetables during and after harvest.

PRESERVING AND PROCESSING OF FOODS

The simple procedures of heating or drying have long been used to preserve food for later consumption. Modern food technology produces a greater variety of more nutritious and flavorful products by combining food-grade chemicals with these preservation procedures. Heating is still one of the most important treatments. Adding acids and salts during the thermal processing of foods for canning results in a more palatable product and helps maintain quality during normal distribution and storage. Stabilizers and emulsifiers are used in canning soups, gravies, and other items that would require considerable time to prepare in the home.

Smoking, which is usually accompanied by heating, is another traditional method of processing foods. Various chemicals in smoke, including carbonyls and organic acids, exert a mild preservative action. Sodium nitrite is usually added to inhibit growth of bacteria that cause botulism as well as to stabilize color. During heating and storage, some of the nitrite may react with other food constituents to produce extremely small amounts of nitrosamines, which are carcinogenic in rats. A recent study indicates that nitrites themselves may be carcinogenic in rats. A concerted effort is underway on the nitrite problem in private and government laboratories. Meanwhile, sodium nitrite is still used in reduced amounts.

Additives are frequently used to prevent the degradation of frozen foods from the reaction of enzymes during storage. A mild heat treatment (blanching) inhibits enzymes in vegetables, but blanching affects the flavor and texture of fruits.   Therefore, enzymes are controlled in most frozen fruits by adding sulfites and ascorbic acid or by packing fruits in sugar.

Drying, probably the oldest food preservation method, preserves food by reducing its water content below the minimum for growth of micro-organisms.   However, some degradation can occur through enzymatic or other chemical reactions.   Sulfur dioxide or bisulfites are commonly used to minimize the degradation.   Fermentation, another traditional method, also uses chemicals, salt being one of the most important.

FORTIFICATION OF FOODS WITH MICRONUTRIENTS

Adding nutrients to foods can help maintain or improve the quality of a diet.   Various terms have been used to describe the process, but the term "fortification" is considered to be the most appropriate.   Fortification is not a new concept; for example, iodine (in the form of iodide) was added to salt in the 19th century in South America.

The nutrients commonly added to food are vitamins, minerals, and, less frequently, amino acids.   The role of essential micronutrients is not completely understood, although the larger gaps in our knowledge of the functions of the minerals required for optimum nutrition are being filled in.   In the future, more trace minerals may be included in fortification programs

As a nutrition improvement method, fortification can be effective where enough food is available but where one or more nutrients is not present in sufficient amounts.   Fortification may be called for in areas where soils are deficient in certain minerals and so contribute to dietary deficiencies.   For example, fluoride is added to water to reduce dental caries, and iodide to salt to control goiter.   Another situation where forti-fication is appropriate is where a population does not have access to or does not consume enough food to provide needed nutrients.   In such circum-stances, long-range approaches are required to meet basic food needs, but a fortification program, if feasible, can quickly introduce certain missing nutrients—iron and vitamin A, for example.   Cereal grain products are particularly appropriate foods to consider fortifying because they provide about 25% of the calories consumed in the United States (and a much higher percentage in the developing nations).

Fortification can be a valuable tool for improving nutrition and health. Its further application depends on a better understanding of nutrition, and on improved fortification technologies to ensure that the added nutrient is uniformly incorporated and remains stable during distribution. At the same time, the sensitivity of a small proportion of the population to increased intakes of certain nutrients—for example, iron—must be considered.

Recommendations

We recommend that:

1. Greater efforts be made to expand fortification of selected foods in those situations where studies have indicated both the need and the feasibility.
2. Fundamental nutrition research be expanded, especially with respect to trace minerals, to provide basic guidance for fortification of foods.
3. Fortification technologies—for example, for adding iron to various foods—be improved.

FABRICATED FOODS

Fabricated foods are made by combining natural and synthetic components to achieve certain nutritional, sensory, and stability characteristics. Four general types of fabricated foods are available: (1) special dietary foods that constitute the entire diet, (2) foods that replace entire meals, (3) foods that imitate traditional foods, and (4) minor foods that ordinarily provide a relatively small percentage of the calories in the diet.

In the industrialized countries, social and economic pressures have led to changes in traditional methods of food production and consumption. Fabricated foods are a response to some of these pressures. First, fabricated foods can be made from a variety of materials that are essentially equivalent nutritionally, making substitutions possible. Second, they can be produced with a consistent composition. Third, they can be produced for people with specific metabolic problems.

Fabricated foods are acceptable only when the consumer is satisfied with their texture, color, taste, and appearance. However, it can be difficult to duplicate the sensory characteristics of traditional foods. There is also the risk that a fabricated food will replace a superior food from natural sources. Thus as the use of fabricated foods increases, the benefits and risks will have to be compared to those of other alternatives.

Recommendations

We recommend that:

1. Methodology be developed to measure and characterize the chemical, physical, and functional properties of food materials, and the role of these properties in the overall quality, ease of processing, and stability of foods.
2. Further studies be undertaken of the nutritional quality and safety of fabricated foods.
3. The benefits, risks, and costs of using fabricated food be evaluated.

SPECIAL DIETARY FOODS

Nutritionists can now formulate foods and diets to satisfy the special needs of various population groups.   For example, special diets are available for sufferers of phenylketonuria, a condition in which the body can metabolize only limited amounts of phenylalanine, an essential amino acid.   Other dietary problems, including weight control, dental caries, and hypertension, call for foods with reduced amounts of sugar or salt.   Because "sweet" and "salty" are basic taste sensations, it is desirable to have nonnutritive sweeteners and substitutes for ordinary salt (sodium chloride).

Nonnutritive sweeteners meet two popular demands—the desire to reduce caloric intake by replacing sugar in the diet, and the need for a substitute for sugar in the diet of diabetics.   The only sweetener now approved by FDA is saccharin, and its safety is the subject of intense national and Congressional debate.

A salt substitute is needed to help treat hypertension, because it can be extremely difficult for a hypertensive to reduce intake of sodium salts.   In many sections of the country, even the water used in cooking may provide a significant amount of sodium salts.   Many common foods also contain substantial quantities.   A number of substitutes have been developed for use by persons on restricted diets, but none duplicate the way ordinary salt affects flavor.

Recommendation

We recommend that:

The search for improved substitutes for sugar and salt be intensified.

UNCONVENTIONAL SOURCES OF FOOD
AND FEED

World food and feed shortages have encouraged special efforts to develop unconventional foods that are nutritious, acceptable, and inexpensive. In recent years, emphasis has centered on single-cell proteins (SCP) and on conversion of waste materials. There has also been some interest in synthetic fats and carbohydrates.

Various attempts have been made to use SCP as a major protein source. Yeast has received the greatest attention. Processes using crude oil fractions, alcohols, and waste materials as raw materials are in or nearing commercial production in the United States and several other countries. Production costs are strongly influenced by raw material costs.

Clinical testing has established that SCP, when appropriately processed, is nutritious and safe for humans and animals. Its major nutritional limitations are its relatively high level of nucleic acid and low level of two essential amino acids. The nucleic acid problem has led to the development of special mutant strains of microorganisms that excrete their nucleic acid when the temperature shifts. These strains are probably used in some of the commercial processes now being developed.

To date, SCP has been used primarily as an animal feed supplement. Its main competitor, soybean meal, costs less. However, an upgraded SCP has potential use in human foods or in such special markets as pet or fish foods, assuming costs are reasonable.

Many categories of waste—corn stalk, cereal straw, and wood wastes, for example—are valuable sources of feed or food as well as of energy, soil conditioners, and fertilizers. A significant part of the protein in animal feeds now consists of crude agricultural by-products formerly considered wastes. Conversion of agricultural wastes to human foods is limited. Questions of safety must be resolved and consumer acceptance achieved before consumption will increase.

Synthesis of fats and carbohydrates may provide energy-supplying foods. Likely starting materials for fats are hydrocarbons and such gases as carbon monoxide and methane, all of which are derived from petroleum. However, more research is needed on the toxicology and nutritional value of synthetic fats and on production costs. Consumer acceptance does not seem to be a problem. However, the present outlook for synthetic carbohydrates is not promising.

Recommendation

We recommend that:

Efforts continue to produce safe feed and food more economically from unconventional sources.

ASSURING THE WISE USE OF CHEMICALS

For thousands of years, people have used chemicals to preserve or improve the flavor of their food. The problem has been, and continues to be, to assure that chemicals are used wisely. There are many more safeguards now than in the past, the major one being the evaluation of safety required by the Food, Drug, and Cosmetic Act. Safety evaluation is more rigorous than ever before. Sometimes the results raise doubts about the substances used in the American food system and how they are regulated, as the saccharin controversy illustrates. But virtually every substance is hazardous if the intake level is high enough. As the search continues for more subtle and infrequent effects, occasional disconcerting findings can be expected.

Present U.S. food laws provide that as long as risk is not a significant factor, utility is best judged in the marketplace where cost/benefit considerations are paramount. Risk, however, is narrowly construed. Comparative risks are not considered, except informally, because there is no statutory authority for their consideration. People are usually highly reluctant to accept any significant, definable, involuntary risks in their food supply. Unfortunately, some degree of risk is always present, so a shift to an apparently less hazardous alternative may only mean a shift to one whose risks have been evaluated less carefully. For example, DDT was replaced with demonstrably more toxic, though less persistent, pesticides.

The ability to detect the presence of chemicals in food is vastly ahead of the ability to detect and interpret their toxicological effects. Much of our current knowledge of toxicity comes from animal tests. However, such tests differ from real-life human exposure in several important respects, including exposure level, species differences, and the genetic homogeneity of inbred laboratory animals in contrast to the genetic diversity of the human species.

Animal feeding tests in which the result is damaged tissues or dead animals demonstrate the existence of potential risk, but may provide no understanding of how risk may occur or of its relevance to human exposure. Greater confidence in safety testing will only come from studies that seek an understanding of what happens to a chemical and its metabolic products in the human body. Analytical chemistry is critical in such studies. To date, advances in analytical cnemistry have enhanced sensitivity, but now greater simplicity and speed are required. To be most useful, metabolic studies in humans and comparative metabolic studies in humans and other species must be conducted at human intake levels as well as some levels above normal intake to produce a safety margin.

Recommendations

We recommend that:

1. Government, universities, industry, the American Chemical Society, and other professional groups continue efforts to increase public awareness and understanding of the benefits, risks, and costs in using chemicals in the food system.
2. The comparison of benefits and risks, or at the minimum, the comprehensive comparison among risks, be an explicit part of the regulatory process.
3. Studies at a wide range of intake levels be undertaken to determine the metabolic fate in humans and animals of chemicals used in the food system.

Production of Food

The figures of the farmer, shepherd, and fisherman pervade mythology and history; their activities are symbolic of mankind's major sources of food, the cultivation and harvest of crops, animal products, and fish. Men have always attempted to formulate practices that would improve their yields from these basic resources. Thus, long before the chemistry of crops and soils was understood, farmers used natural materials such as manure, compost, leaves, wood ash, saltpeter, and guano as fertilizers. Use of potassium fertilizer in the form of manure, potassium nitrate, and ashes was recorded in 300 B.C. Recognizing the need to control weeds and other pests, early farmers also applied a variety of chemicals for that purpose; the selection was limited to those that were readily available in nature or could be derived easily from metals and their salts.

While these primitive uses of agricultural chemicals were often based on empirical correlations, they provided some of the early glimmerings of scientific understanding. The development of science and the development of agriculture are closely linked, especially with regard to chemistry, although the resources and techniques of contemporary biology, biochemistry, and allied scientific disciplines are also an integral part of food production technology.

U.S. agricultural productivity has increased dramatically since the end of World War II. In the past few years, the increases have not been so great, suggesting the need for new strategies to increase food production. However, the purely quantitative increases in production must be accompanied by attention to the nutritional quality of the product and by increases in the efficiency with which plant, animal, or fish nutrients are converted to food. These developments will require a greater understanding of the biochemistry of the production system and how to manipulate those factors that limit productivity. For example, effective chemical regulators or new technologies that will improve the efficiency of photosynthesis or the fixation of nitrogen by crops can be achieved only by increased biochemical knowledge.

Any new technologies must be introduced with adequate safeguards to avert undesirable ecological changes. It is also important to ensure that new technologies can be assimilated smoothly into the existing social and economic structure. All the major elements in the food production system have felt the impact of technological change. Generally, the changes have been beneficial. With careful attention to all their implications, future changes should be equally beneficial.

CROP PRODUCTION

Widespread use of fertilizers and pesticides has been a major factor in the increased crop productivity noted in the past few decades. Other factors have also been involved, including improved management, better plant varieties and hybrids developed through plant breeding, and mechanization. This increased productivity has meant that while the population of the United States rose from 151 million to 216.4 million between 1950 and 1978, the cropland harvested fluctuated between 286 and 338 million acres (1). During this period, U.S. exports of agricultural commodities, primarily plant products, have increased more rapidly than imports.

Fertilizers

Plants need about 20 chemical elements for growth. Those needed in the greatest amounts—nitrogen, phosphorous, and potassium—are termed major nutrients. Plants also need smaller amounts of calcium, magnesium, and sulfur, the so-called secondary nutrients, as well as minute amounts of other elements called micronutrients.

Use of chemical fertilizers in the United States has expanded fourfold since 1950, and today food production is 30–40% higher than it would otherwise be. For maximum effectiveness, however, fertilizers should be a part of a carefully planned system that includes proper management of all controllable factors involved in productivity. The successful development of fertilizer production technology and its efficient application in modern well-engineered plants have held price increases for fertilizer materials below those of other major farm inputs. For example, from 1950 to 1976, prices of fertilizer materials increased less than 50% as compared with increases of about 400% for farm machinery, 400% for farm wage rates, and 600% for farm real estate *(2)*.

PRODUCTION   About half the fertilizer used in the United States is applied as mixtures generally consisting of more than one major nutrient, and manufactured from two or more intermediates.   Approximately 45% by weight are bulk blends (physically mixed granular materials), 37% are granular compounds containing more than one nutrient, and 18% are liquids *(3)*.   The mixtures are normally compounded in small plants located in the market areas, with the fertilizer materials brought in from larger primary production units located at raw material sources or along main transportation routes.

Detailed information about the chemistry and technology of fertilizer production has been published and the Tennessee Valley Authority has compiled bibliographies for more in-depth coverage *(4)*.

Nitrogen.   Practically all nitrogen fertilizer is derived from ammonia (NH_3).   About 37% of the ammonia fertilizer used in the United States is applied directly to cropland, and the remainder is used to produce urea, ammonium nitrate, ammonium sulfate, and compounds containing more than one nutrient.   The trend is toward the increasing use of urea.

More than 95% of U.S. ammonia is made from natural gas and nitrogen from the atmosphere by a series of reactions that take place at elevated pressures and temperatures in the presence of catalysts.   In the reactions, nitrogen combines with hydrogen in natural gas.   About 36,000 cubic feet of natural gas are required to make a ton of ammonia,.two-thirds of it to suppy hydrogen and one-third to supply process energy.   (See the accompanying box on the synthesis of nitrogen fertilizers.)

Urea is produced by high-pressure, high-temperature reaction of ammonia and carbon dioxide.   Under typical conditions, only about 50% of the materials react, so that unreacted materials must be separated and recycled.   Processes differ mainly in the manner of separation and recycling of unreacted materials.   The typical product from a urea plant is a 75% solution, which usually is concentrated to about 99% and then solidified into prilled or granular forms.   More recently, the 75% solution is being physically mixed with 83% ammonium nitrate solution to produce a liquid containing 28–32% nitrogen.

Ammonium nitrate is made by the neutralization of nitric acid with ammonia.   Normally, the product is a solution of 80–90% concentration, which is used to produce urea-ammonium nitrate solutions or evaporated to near dryness and either prilled or granulated.

Synthesis of Nitrogen Fertilizers

The technology for manufacturing ammonia was developed by German scientists shortly before World War I.   The raw materials for the process are nitrogen from the atmosphere and hydrogen, which can be obtained from many sources.   In the United States, more than 95% of ammonia is made using hydrogen from natural gas, which is largely methane, in a series of reactions that take place at high temperatures and pressures in the presence of catalysts.   The methane is broken down into hydrogen in the following reaction:

$$CH_4 + H_2O \longrightarrow 3H_2 + CO$$

The hydrogen and nitrogen from the atmosphere react, again at high temperatures and pressures in the presence of catalysts, to produce ammonia:

$$3H_2 + N_2 \longrightarrow 2NH_3$$

The ammonia can be used directly as a fertilizer or it can be reacted with a number of acids to produce various forms of fertilizer materials, including:

$$2NH_3 + H_2SO_4 \longrightarrow (NH_4)_2SO_4 \qquad \text{ammonium sulfate}$$

$$NH_3 + HNO_3 \longrightarrow NH_4NO_3 \qquad \text{ammonium nitrate}$$

$$3NH_3 + H_3PO_4 \longrightarrow (NH_4)_3PO_4 \qquad \text{ammonium phosphate}$$

At high temperatures and pressures, urea fertilizer can be produced by reacting ammonia and carbon dioxide in a two-step process. In the first step:

$$2NH_3 + CO_2 \longrightarrow NH_2COONH_4 \qquad \text{ammonium carbamate}$$

The product is then dehydrated to form urea:

$$NH_2COONH_4 \longrightarrow NH_2CONH_2 + H_2O \qquad \text{urea}$$

Phosphate.   Essentially all phosphorus fertilizer is derived from phosphate rock.   About three-fourths of U.S. phosphate fertilizer comes from rock mined and processed in central Florida.   After being processed, the material is converted into fertilizer products or such intermediates as phosphoric acid, which supplies about 85% of fertilizer phosphate in the United States. Treating phosphate rock with sulfuric acid produces liquid phosphoric acid and by-product calcium sulfate dihydrate, a solid that is removed by filtration. The concentration of the acid, expressed as phosphorous pentoxide (P_2O_5), is 27–30%. The acid may be evaporated before use to a concentration of 40–42% P_2O_5 or 52–54% P_2O_5. Reacted with ammonia, the acid produces such fertilizer compounds as solid monoammonium phosphate ($NH_4H_2PO_4$), solid diammonium phosphate ($(NH_4)_2HPO_4$), or solutions of ammonium phosphates. Granular diammonium phosphate comprises about two-thirds of all ammonium phosphate fertilizer marketed in the United States.

Triple superphosphate is produced by reacting phosphoric acid with phosphate rock in proportions such that about 70% of the phosphorus is derived from the acid. The principal phosphate in triple superphosphate is monocalcium phosphate. The granular form gradually is replacing the powder form.

Potassium. About 90% of potassium fertilizer or potash is derived from potassium chloride, and most of the remainder from potassium sulfate. Potassium chloride is recovered by physical processing of minerals derived mostly from underground salt deposits. Potassium fertilizers may be applied directly or as mixtures.

CONSUMPTION TRENDS   Nitrogen. In the year ending June 30, 1979, about 10.6 million short tons of nitrogen fertilizer, expressed as the element nitrogen (N), were consumed in the United States. For the near term, demand for nitrogen fertilizers can be projected to grow at about a 5% annual compound rate *(5)*. It appears that if new plants come into operation as scheduled, the supply of fertilizer nitrogen will be adequate to meet demand into the early 1980s.

However, some factors cloud the market outlook. Although less than 3% of U.S. natural gas consumption goes to produce nitrogen fertilizers, production must compete for the limited reserves (the known deposits that can be removed from the earth at reasonable cost). Currently, reserves are estimated to be about 200 trillion cubic feet, or ten years' supply at present use rates *(6)*. Until 1978, only the price of interstate gas was regulated so that it was much cheaper than unregulated intrastate gas. Hence about half of U.S. ammonia capacity is located on interstate gas supply points. During the few years preceding 1978, interstate gas supplies were not always adequate to meet demand during winter months, and ammonia production was sometimes curtailed. Gas curtailments in the 1976–77 winter led to the loss of more than 700,000 tons of ammonia production.

The Natural Gas Bill of 1978 regulates the price of all natural gas and allows the price to rise gradually until 1985, at which time all controls will be lifted.   This legislation should increase supplies of natural gas.   However, natural gas eventually will be in short supply, and the price will increase substantially.   As recently as ten years ago, gas was available to producers of ammonia on the Gulf Coast for as little as 20 cents per thousand cubic feet, contributing about $7 to the cost ($12 total) of a ton of ammonia.   The price of unregulated or intrastate gas in 1979 was $2 or more per thousand cubic feet.   At this price, gas costs about $70 per ton of ammonia, and total production cost is about $120.   Gas prices may rise to the range of $3 to $4 per thousand cubic feet during the 1980s.   At these prices, the equivalent cost of ammonia production in 1979 would be $155 to $190 per ton.

In view of the natural gas supply and related cost problem, the following alternatives can be considered as a means of supplying nitrogen fertilizers:

- *Import nitrogen fertilizer.*   Because of the uncertainty of natural gas supplies in the United States, some nearby countries are expanding their ammonia capacity with the idea of exporting in the U.S. market. Three new ammonia plants in western Canada are scheduled to come into production by 1980, and new units are scheduled to start up in Trinidad and in Mexico.   However, dependency on foreign sources for an essential input to the nation's agricultural system is a questionable course of action.

- *Use oil as a raw material.*   United States oil reserves, like natural gas, are limited.   The nation's oil supply could be sharply reduced within the next 30 years *(7)*, and dependency on foreign oil imports is already creating severe balance of payment problems.   Like natural gas prices, oil prices can be expected to climb rapidly as supplies are reduced, assuming price controls are removed.

- *Use coal as a raw material.*   At present rates of use, U.S. coal could supply the hydrogen and energy for ammonia production for hundreds of years.   Ammonia is being produced overseas from coal in several relatively inefficient plants based on technology developed prior to World War II.   Better technology is needed for use of U.S. coals.   This alternative seems reasonable and attractive, and if technology development is emphasized, could be implemented before gas is more expensive and in short supply.

- *Electrolyze water to supply hydrogen.*   This process is used at locations such as Norway with surplus energy from hydroelectric plants.   However, total energy requirements for this route are about twice those for direct production from coal.   Therefore, electrolysis is attractive only where energy supply is abundant; technological breakthroughs may improve the efficiency of the system.

- *Legislate the highest priority for use of natural gas in fertilizer production, while requiring other users to switch to alternative energy sources.* By assuring adequate nitrogen fertilizer supplies for the nation until well into the next century, such legislation would provide more time for development of alternative technology.

- *Use manures or other natural organic materials as fertilizers.* Such use has the potential to conserve both materials and energy. However, natural fertilizers contain an average of only 2% nitrogen, versus 82% for ammonia and 46% for urea. In addition, in-depth investigations have revealed that the U.S. requirement for nitrogen fertilizers cannot be met through this route. For example, if the 7 million tons of dry sewage sludge generated annually were added to 170 million tons of recoverable dried animal manure, the total would provide approximately 1.8 million tons of nitrogen or about 17% of the current annual nitrogen demand *(8)*. Also, the transportation and handling costs for manures and sewage sludge are much higher than for urea and ammonia. Furthermore, toxic metals and other contaminants in manures and sludge could be troublesome.

- *Rely on legume crops.* During the early days of U.S. agriculture, obtaining fertilizer nitrogen from legumes by crop rotation was a general practice. Although still useful in some circumstances, this procedure can no longer serve as a major source of fertilizer nitrogen because not enough agricultural land is available to permit rotation of agricultural use with legume cover.

- *Develop better means for direct fixation of nitrogen from the air by chemical or biological means.* Research on this alternative (discussed later in this chapter) is in the very basic stages, and early breakthroughs are not likely.

Thus it appears that beyond the early 1980s, the U.S. supply of nitrogen fertilizer will depend on the availability of natural gas. If the use of gas by ammonia plants continues to be restricted, then more nitrogen fertilizer will have to be imported until new technology can be developed to produce it from raw materials other than natural gas.

Phosphate. In the year ending June 30, 1979, total consumption of phosphate fertilizer in the United States was 5.5 million short tons P_2O_5, with about 67% as ammonium phosphate and other compounds, 28% as triple superphosphate, and 5% as single superphosphate. Supplies appear to be adequate to meet demand through 1980. Any upward shift in demand probably could be met by an increase in operating rates of existing plants. Also, some obsolete plants not now in operation could probably be brought back into production if demand should unexpectedly exceed supply, with resultant increases in prices.

Phosphate resources in the United States are adequate for hundreds of years *(9)*. However, known supplies of the best quality material are being depleted. In the near future, many of the major Florida producers will begin mining in areas south of their present operations, and ore quality problems will increase—particularly problems associated with increasing proportions of dolomite. In addition to the reserves in Florida, other reserves that will become increasingly important are located in North Carolina and Idaho. New technology must be developed to use this marginal quality phosphate efficiently.

Most major U.S. phosphate reserves contain uranium. The process used to produce 85% of U.S. phosphate fertilizers concentrates most of that uranium in the phosphoric acid intermediate containing 27–30% P_2O_5. Each ton of acid P_2O_5 made from typical Florida phosphate rock contains about one pound of uranium oxide (yellow cake). Its net energy value is estimated to be more than 100 million Btu's, more than ten times the energy requirement for producing the ton of P_2O_5 *(10)*.

New solvent extraction technology can recover yellow cake at a cost of approximately $20 per pound *(11)*. On the open market, the price for yellow cake in 1979 was over $40 per pound. Therefore, from the economic and energy conservation viewpoints, recovery of uranium may be attractive. At least two U.S. companies have announced plans to construct facilities for recovery of uranium from phosphoric acid.

Potassium. Total consumption of potassium fertilizer in the United States was about 6.2 million short tons of potassium oxide (K_2O) for the year ending June 30, 1979. About 80% of the potassium fertilizer used in the United States is derived from Saskatchewan. Potash resources of the world are immense, and at present rates of consumption it seems unlikely that any shortage will develop.

Consumption of potassium fertilizer is predicted to grow to about 6.5 million tons of K_2O by 1980 *(12)*. In 1977, U.S. capacity was about 3.5 million tons of K_2O per year, with Canadian capacity reported at about 11 million tons per year. The combined U.S. and Canadian capacity far exceeds the combined demand for the two countries, and this will continue to be so for the foreseeable future. Even though the United States has significant reserves of potassium, Canadian reserves are much more extensive, of better quality, and can therefore supply a product at lower cost than can be produced in the United States. However, Canadian producers have been under a quota system and are now facing a partial nationalization of the industry. Hence many potash producers are looking for alternative sources of supply.   This will result in increased interest in lower grade U.S. material or in material for which the economics of mining has not been attractive previously.

TECHNOLOGY TRENDS   Several major trends in fertilizer technology have become apparent in the past few decades:

- *Relative consumption of urea and ammonium phosphates has increased.*   The proportion of U.S. nitrogen fertilizer consumed as solid urea has doubled and is now about equal to that of solid ammonium nitrate (Figure 2.1) *(13)*.   Since 1970, the proportion of phosphate fertilizers produced as ammonium phosphates has also increased, rising from about 10% in 1960 to about 66% in 1978 (Figure 2.2) *(13)*.   Both urea and ammonium phosphates contain a higher percentage of nutrients than materials they are supplanting, which has contributed to increasing the concentrations of major nutrients in commercial fertilizers from 20% in 1940 to about 45% in 1978.

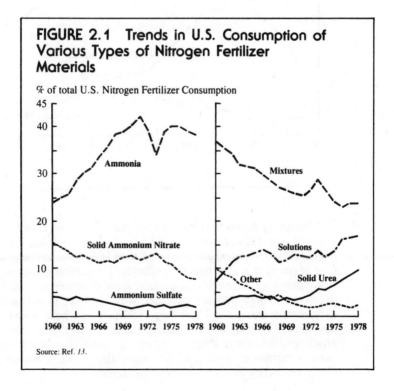

FIGURE 2.1   Trends in U.S. Consumption of Various Types of Nitrogen Fertilizer Materials

% of total U.S. Nitrogen Fertilizer Consumption

Source: Ref. *13*.

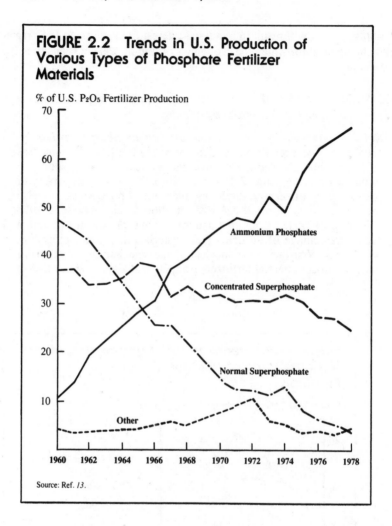

FIGURE 2.2 Trends in U.S. Production of Various Types of Phosphate Fertilizer Materials

% of U.S. P₂O₅ Fertilizer Production

Source: Ref. *13*.

- *Production of liquid fertilizers has increased.* From about 7.7 million tons in 1967, production rose to 15.0 million tons in 1978. Liquids now comprise about 33% of total U.S. fertilizer production. One segment that may become particularly significant is the suspension category, mixtures that contain more fertilizer salts than will dissolve. These salts are stabilized with additives to minimize settling during storage and handling. Suspensions have a number of advantages: they are more concentrated than clear liquids; they can be produced from low quality, less costly raw materials; and they can be used as carriers of some insoluble materials, including certain pesticides, secondary nutrients, and micronutrients.

● *Primary production plants have become larger and increasingly efficient since the 1950s.* Ammonia, phosphoric acid, and granulation plants with outputs of 1,000 tons per day or more are commonplace. During the 1970s, plant size tended to level off at 1,500 tons per day.

PROBLEMS AND ALTERNATIVES   The major problems for chemical fertilizers relate not only to the raw material supplies for nitrogen and phosphate, but also to the low efficiency with which fertilizers applied to crops are used, a shortage of energy resources, environmental concerns, and the special technological needs of developing countries.

Efficiency of use.   Probably only about half of the fertilizer applied is efficiently used by the crop.   Reducing these losses would conserve the raw materials and energy required in production and also decrease the potential environmental damage by fertilizers not used by the crop.

Both nitrogen and potash fertilizers dissolve in any water in the soil and so can move significantly after application, nitrogen to a greater extent than potassium.   Nitrogen in the ammonium form is held tightly to soil particles by ionic attraction.   However, all ammonium or ammonium-producing fertilizers can be converted by certain soil bacteria to nitrate (the process of nitrification).   Nitrate may then be lost from the soil through leaching or through conversion by other bacteria to gases.   Controlling nitrification can increase the efficiency of nitrogen fertilizers.   This can be done by adding to nitrogen fertilizers chemicals that inhibit the activity of the nitrifying bacteria.   Another possibility is products with slow- or controlled-release properties, such as coated products and products that are inherently slow to dissolve in water.   Either approach has been shown to improve nitrogen fertilizer efficiency under conditions such as those occurring after heavy rains, where nitrate is formed and consequently nitrogen is lost.

When urea is applied at the surface, the loss of nitrogen can be significant if the soil holds little moisture.   Coating the urea or mixing it with other fertilizer materials can reduce the losses.

Energy–fertilizer problems.   Production and distribution of fertilizers account for about 600 trillion Btu's per year, which is about 0.6% of the total energy consumed in the United States and about 20% of the energy *(14)* consumed annually by the U.S. agricultural system, including harvesting the crop.   Ammonia production accounts for about 70% of the energy required to produce fertilizers.   Other significant energy uses are associated with production of phosphoric acid; transportation, storage, and handling of raw materials, intermediates, and products; and drying of granular fertilizers, which comprise about 70% of U.S. production.   Most of the granular products require a drying step during processing, which is a major consumer of energy, a major cost, and a contribution to air pollution. Research is underway on granulation processes that require no drying. The chemical heat available from reactions that occur as the raw materials are mixed during granulation should be employed to the fullest extent to drive off water. Most granulation formulations have substantial chemical heat because of the ammonia and mineral acids they contain.

Environmental concerns. The chemical fertilizer industry has made substantial improvements in its operations in recent years to meet pollution regulations. Probably the most important opportunities for further improvement relate to (1) finding more effective ways to recycle effluents so that they are not discharged to the environment and (2) decreasing energy usage.

The use of chemical fertilizers also has the potential for damaging the environment. One fear is that the continuing intensive use of nitrogen fertilizers may lead to increased nitrate levels in surface and ground waters. This, in turn, could lead to increased eutrophication of lakes and streams and to health hazards to livestock and humans through higher levels of nitrate in drinking water. Studies have shown that little nitrogen is lost by leaching in the presence of an actively growing crop. However, more data are needed on both leaching and runoff of nitrogen from field soils. Nutrient balance studies are difficult to carry out under large-scale field conditions, but available results show that more nitrogen is added to surface waters from precipitation than from drainage of fertilized agricultural fields *(15)*.

The traces of several heavy metals in phosphate rock are another concern. Generally, most of the concern is about the effects of uranium and cadmium that could enter the food cycle with phosphate fertilizers. The recovery of uranium from phosphoric acid for economic reasons could ultimately remove most of this element from phosphate fertilizer products. Fortunately for the United States, Florida phosphate rock has much less cadmium than any of the other commercial phosphates in use. The ultimate fate of cadmium in fertilizer materials is under continuing study, but results to date indicate that plant uptake is not significant at the rates at which phosphorus is usually applied under field conditions *(16)*.

Special technology for developing countries. Fertilizer technology developed specifically for developing countries is a high priority item. Most existing technology originated in industrialized nations in the temperate zone and is not necessarily appropriate in tropical and subtropical areas. New fertilizers based on local raw materials are needed, and adapting existing fertilizers to tropical conditions could make them at least as efficient as they are in temperate zones.   In addition to changing methods of providing the major nutrients, new approaches to furnishing secondary nutrients and micronutrients are also needed.

United States developments in chemical fertilizers are being put to work by the International Fertilizer Development Center, an independent, nonprofit public corporation recently set up adjacent to TVA's National Fertilizer Development Center in Muscle Shoals, Alabama.   Its mission is the development of new and improved fertilizers, along with methods to distribute and use them, to help increase the output of tropical and subtropical agriculture.

Pest Control

At every stage of production, the yields and quality of food crops are reduced by the relentless activities of pests and predators.   Newly planted seeds may be consumed by insects, rodents, or birds, or damaged by disease.   Growing plants must contend with weeds, disease, nematodes, and soil-borne insects.   Maturing fruits and grains may be infested by insects, rotted by fungi, or consumed or damaged by birds or other animals.   Pests are equally detrimental in other segments of the food system. After harvest, insects, microorganisms, or rodents may destroy crops that are being stored or shipped.   Rodents, for example, contaminate vast amounts of food with excreta, in addition to the food they actually consume.   Cattle herds infested with insects may curtail meat production, and the quality, flavor, and quantity of dairy products are adversely affected by microorganisms, weeds, and insects.   There are many weeds that are poisonous— and sometimes even lethal—to humans, livestock, and wildlife.

The problems of pests and predators are as old as man's struggle to produce and store food.   Before 1850, over 20 chemicals were in use to control pests in European agriculture *(17)*.   Burning sulfur, which produces sulfur dioxide, was advocated as a fumigant by several Roman and Arab agricultural writers, and the element itself was suggested as a fungicide in 1779.   The use of another fungicide, copper sulfate, for treating seed was first mentioned in 1761.   Bordeaux mixture, a complex of copper sulfate and slaked lime, was introduced in 1885 to control mildew on grape vine.   Although Bordeaux mixture is still in use, less corrosive copper complexes are now preferred.   Likewise, polysulfide derivatives, including "lime sulfur," made by boiling sulfur with a suspension of lime, are still used to control fungal diseases.

TYPES OF PESTICIDES Almost all of the pesticides used in the United States today are organic chemicals synthesized from petrochemicals. The major classes of crop-protection chemicals include herbicides, insecticides, fungicides, nematicides, miticides, acaricides, viricides, bactericides, and fumigants. Insect attractants and repellents also have their roles in pest management. Nor are repellents confined to insects. Many birds and mammals interfere with food production, and chemical treatment may effectively deter such predators.

For regulatory purposes, the term "pesticide" includes many types of chemicals used in food production in addition to those lethal to pests. Examples include defoliants used in harvesting cotton and other crops, desiccants such as those used for preharvest kill of potato virus, plant growth regulators, soil sterilants, and other chemicals within the very broad area that falls under the jurisdiction of the Federal Insecticide, Fungicide, and Rodenticide Act of 1947, as amended in 1972 and 1975. The amendments broadened the implications of the Act by regulating all aspects of pesticide use, including application, storage, and disposal.

A recent survey lists 50 organic pesticides of current and future importance (Table 2.1) *(18)*, many of which contribute to increasing the efficiency of agricultural production. The largest class of pesticides is the herbicides. The need to control weeds is an ancient problem, as evidenced by use of hand-weeding tools as early as 6000 B.C. Since then, there has been continuous improvement in machinery and cultural practices, but only recently has there been a substantial decrease in the human effort required to control weeds. It has been known since biblical times that certain inorganic chemicals kill weeds. By the early 20th century, sulfuric acid, ammonium sulfamate, sodium arsenite, and other inorganic chemicals were used for this purpose. Sodium chlorate is still in general use for those situations where all vegetation must be destroyed.

The application of weed-control chemicals on a large scale is a phenomenon of the mid-20th century. Synthetic organic herbicides were introduced in the 1930s with the dinitroalkyl phenols, although the insecticidal value of dinitrophenols had been recognized in the 19th century. This group of compounds is toxic not only to insects and plants but to mammals as well. They remain in use to the present as selective herbicides.

Selective herbicides may be applied to a growing crop, where they kill weeds without damage to the crop. Their selectivity may depend on physical or biochemical differences between the weeds and the cultivated plants. Sulfuric acid provides one of the earliest examples of physical differences. Cereal crops (monocotyledons) have narrow waxy leaves, whereas the weeds (dicotyledons) are broad-leaved. Applied as a spray, sulfuric acid runs off the narrow leaves of the cereal plant, but the broad leaves of the weed receive and retain more of the acid.

TABLE 2.1   Some Organic Pesticides of Major Current and Future Economic Importance

Pesticide	Approximate Year of Introduction	Use
Methyl bromide	1932	Space and soil fumigant
2,4-D	1942	Postemergence herbicide
Dichloropropane-dichloro- propene mixture	1942	Soil fumigant
Zineb	1943	Foliar fungicide
Dinoseb	1945	Pre- and postemergence herbi- cide
Methyl parathion	1947	Foliar insecticide
Captan	1949	Foliar fungicide
Parathion	1949	Foliar insecticide
Maneb	1950	Foliar fungicide
Malathion	1950	Foliar and premise insecticide
Diazinon	1952	Soil and foliar insecticide
Dalapon	1953	Postemergence herbicide
Azinphosmethyl	1953	Foliar insecticide
Phorate	1954	Soil and systemic insecticide
Diuron	1954	Preemergence herbicide
EPTC	1954	Preemergence herbicide
Dicofol	1955	Acaricide
Dibromochloropropane	1955	Soil fumigant
Disulfoton	1956	Soil and systemic insecticide
Carbaryl	1956	Foliar insecticide
Simazine	1956	Preemergence herbicide
Atrazine	1958	Preemergence herbicide
Chloramben	1958	Preemergence herbicide
Paraquat	1958	Postemergence herbicide
Fenitrothion	1959	Foliar insecticide
Trifluralin	1960	Preemergence herbicide
Linuron	1960	Pre- and postemergence herbi- cide
Fluometuron	1960	Preemergence herbicide
Triallate	1961	Preemergence herbicide
Butylate	1962	Preemergence herbicide
Pyrazon	1962	Pre- and postemergence herbi- cide
Chlorthalonil	1963	Foliar fungicide
Picloram	1963	Postemergence herbicide
Bromacil	1963	Preemergence herbicide and soil sterilant
Aldicarb	1965	Nematicide and systemic insecticide
Dicamba	1965	Postemergence herbicide
Propachlor	1965	Preemergence herbicide
Chlorpyrifos	1965	Soil, foliar, and premise insecticide

(continued)

TABLE 2.1   Some Organic Pesticides of Major Current and Future Economic Importance

Pesticide	Approximate Year of Introduction	Use
Chlordimeform	1966	Foliar insecticide and acaricide
Alachlor	1966	Preemergence herbicide
Methomyl	1966	Foliar insecticide
Carbofuran	1967	Soil insecticide and nematicide
Benomyl	1967	Systemic foliar fungicide
Cyanazine	1968	Preemergence herbicide
Cyhexatin	1968	Acaricide
Bentazon	1968	Postemergence herbicide
Tridemorph	1969	Systemic foliar fungicide
Orthene	1971	Foliar insecticide
Metribuzin	1971	Preemergence herbicide
Glyphosate	1973	Postemergence herbicide

Source: Ref. *18.*

Selectivity based on biochemical differences is now an established feature of many common synthetic herbicides. Since the halogenated phenoxy herbicides such as 2,4-D (2,4-dichlorophenoxyacetic acid) were introduced in the 1950s, their production and use have grown, and many other new synthetic herbicides have been marketed. Phenoxy herbicides can be used to kill broad-leaved weeds in cereal crops because they act selectively against the weeds. The mechanisms of selectivity are not always understood. In the case of atrazine (2-chloro-4-ethylamino-6-isopropylamino-s-triazine), a pre-emergent herbicide, selectivity depends on the ability of corn plants to detoxify the herbicide by hydrolysis, dealkylation, and inactivation through combination with glutathione.

Insecticides are another major group of pesticides. Natural products, particularly the pyrethroids, have been used as insecticides since the discovery in the early 1800s that certain Caucasian tribes used the flowers of a plant as an insecticide. The production of pyrethroids from natural sources increased through the 19th and 20th centuries, until the manufacture of synthetic pyrethroids commenced in the 1950s. The new synthetic pyrethroids developed in recent years show promise as highly effective insecticides that retain biological activity for a longer period than the natural compounds, which are readily decomposed in sunlight. The synthetic pyrethroids, which are relatively nontoxic, may help take the place of several chlorine-containing insecticides that the U.S. Environmental Protection Agency (EPA) has banned on such major crops as cotton. Other classes of synthetic insecticides, such as the organophosphates and carbamates, can still be used for many insect control problems.

Fumigants are another type of pesticide, one of growing importance because much of the world's food is stored for long periods before consumption. There is increasing interest in maintaining adequate food reserves to meet emergencies. Fumigants and other chemicals used to protect stored food from the depredations of pests are not entirely satisfactory. Recent studies of the toxicological effects of several organic chlorine- and bromine-containing fumigants, including some used on fruits and grains, have shown that the fumigants may adversely affect exposed animals or animals fed on a diet of treated feeds.

DEVELOPMENT AND PRODUCTION   To develop a new pesticide, industry must invest considerable time and capital, not only to synthesize and discover chemicals with significant biological activity but also to undertake the research needed to ensure that the product can meet registration requirements. For example, $10–15 million, spread over 15–20 years, would be required to bring a new pesticide to the marketplace, according to 1975 data *(19)*.

Before a pesticide can be registered, EPA must determine if it can perform its intended function without "unreasonable adverse effects," which are defined as "any unreasonable risk to man or the environment, taking into account the economic, social, and environmental costs and benefits of the use of any pesticide." The complexities of balancing risks and benefits of new pesticides mean that a registration application requires several years to process.

Preliminary examination of tens of thousands of compounds is usually required for the discovery of a single marketable product. Initial testing in a laboratory or greenhouse indicates type of activity and range of usefulness. Problems such as the potential toxicity of an insecticide to growing plants may be recognized at this stage. Larger-scale synthesis then supplies material for field testing. At this stage, the method of application, dose rate, and timing must be evaluated to achieve optimum effectiveness.

Subsequently, some of the promising compounds are taken to the field for testing against various pest species. Ultimately, a few are selected for development of appropriate formulations. Active pesticide ingredients are not generally sold directly to the user. Instead, they are formulated (a process constituting a technology in itself) into a diluted material that can be applied to the target site efficiently with minimal risk of injury to other organisms or the environment.

Formulations may be wettable powders, emulsifiable concentrates, concentrated solutions, dusts, granules, or a variety of other preparations that can be applied by conventional spray equipment after dilution. To ensure effective application at any site over a wide climatic range, formulations require many additives. For example, stickers cause the active ingredient to adhere to a plant leaf or other surface; surfactants maintain the stability of the suspension or emulsion when it is diluted with water, increase coverage of leaf surfaces, and improve contact with absorption sites on the leaf; and thickeners maintain insoluble chemicals as a uniform suspension or prevent drift of aqueous spray solutions. Efficacy of a formulation must be demonstrated under a variety of conditions because soils, cropping methods, and many other factors vary from location to location.

When initial field trials have been completed, official trials must be undertaken on a larger scale over a longer period. If it appears that the product may be commercially successful and its utility can be demonstrated, the manufacturer must apply for registration. Before a pesticide can be registered, safety tests must have been conducted to evaluate hazards to man, the environment, and nontarget species. Animal toxicity tests may require several years to complete and may start as soon as a pesticide appears to hold promise. In addition to gathering acute and chronic toxicity data, it is necessary to determine the nature and quantity of residues to which consumers might be exposed if the pesticide were to be used on food or feed plants or on animals. The analytical chemist must therefore devise specific and sensitive methods of analysis for the parent compound and any of its transformation products.

If residues might occur on foodstuffs, a tolerance or exemption (if toxicological data indicate the residue is harmless) must be set by EPA. A tolerance, as defined by the Food and Agriculture Organization and the World Health Organization (20), is "the maximum concentration of pesticide residue that is permitted in or on food at a specified stage in the harvesting, storage, transport, marketing, or preparation of food, up to the final point of consumption." The concentration is expressed in parts by weight of the pesticide per million parts by weight of food. EPA bases acceptable tolerance levels on extrapolation to man of animal tests in conjunction with considerations of metabolic data, dietary intake, and probable exposures. The U.S. Department of Agriculture enforces tolerances for meat and poultry, and the Food and Drug Administration (FDA) for all other foods, under provisions of the Federal Food, Drug, and Cosmetic Act as amended.

The chemist plays an important role in determining the environmental fate of the pesticide and the quantitative and qualitative aspects of its metabolism.   This information forms an important part of the data that the manufacturer must present when applying for registration.   Subsequently, a "label" must be approved specifying permitted dose rates, methods of application, and the crops, commodities, and areas where use of the formulation is sanctioned, as well as an ingredient statement and necessary safety precautions.

From the synthesis efforts of the chemist in the laboratory, through pilot plant and large-scale manufacture to the complex process of formulation, the development of a pesticide calls for a considerable investment of time and money.   Field trials and safety data are also extremely costly.   As a result, development of a pesticide is economically feasible only if a sufficiently large market exists, which usually means that it is to be used against important pests on such major crops as cotton, corn, and soybeans. These economic constraints have resulted in notable gaps in the armory of pesticides available to the farmer, especially pesticides for use against unusual pests on major crops as well as against all important pests on economically less significant crops.   Financial or regulatory incentives may be needed to encourage production of pesticides to fill these gaps.   For example, it may be desirable to find new ways of financing the studies needed to establish the efficacy and safety of limited-market pesticides, or to simplify the registration process.

Production.   The spectacular growth of synthetic organic pesticides during the past four decades can be traced to economic and technological progress of the chemical industry throughout Western Europe, Japan, and North America.   Systematic synthesis and screening of complex organic compounds for pesticidal activity began in the 1930s, and World War II saw the introduction of synthetic organic insecticides on a large scale.   Although the war set the stage for the wide use of DDT, restrictions on free exchange of information delayed the widespread use of many new insecticides developed in Europe.   The 1950s witnessed introduction of the carbamates, DDT, and other chlorine-containing organic insecticides.   Use of organochlorine pesticides has declined considerably in the United States in recent years.   Several have been banned because of concern over the buildup of residues in the environment.   The increasing tendency of some insects to develop resistance to the materials, along with the availability of other insecticides, has also contributed to the decline.   Development of organophosphates (introduced in the late 1940s) and carbamates continues and has been recently supplemented by the introduction of synthetic pyrethroids.

Similarly, large-scale production of weed-control chemicals began in the 1950s with the introduction of 2,4-D.   The ingenuity of U.S. industry has produced new and varied types of chemicals that possess biological activity. Among the herbicides, many major families of organic compounds are represented. Important classes of herbicides include the phenoxy, benzoic, and aliphatic acids, carbamates, thiocarbamates, anilides, dinitro-aniline derivatives, triazines, pyridazinones, quaternary pyridylium salts, uracils, and ureas. Organic arsenicals and sodium chlorate represent the major metal-containing herbicides still in use.

The increase in production and sales of pesticides in the United States during the past 25 years (Figures 2.3 and 2.4) *(21)* indicates their acceptance and value. The over 1 billion pounds of pesticides produced annually in the 1970s include products such as fumigants, wood preservatives, nemati-cides, and rodenticides, but the largest proportion is herbicides and insecti-cides. Herbicides dominate in terms of production and sales. In 1977, they represented 49% of U.S. pesticide production. Sales of herbicides represented 62% of the dollars spent on pesticides *(21)*—a 79-fold increase in value since 1951.

There is no evidence of changes in these trends toward increasing usage. Shifting patterns of agriculture and appearance of pest populations that are resistant to chemicals now available will spur development of new materials. Another stimulus is the need to seek replacements for a number of chemicals in current use that may present unacceptable risks to man and the environment. However, the use of major pesticides is expected to continue to increase.

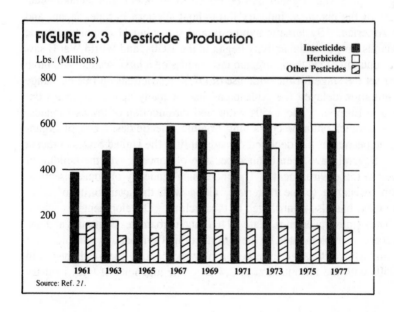

FIGURE 2.3   Pesticide Production

Insecticides ■
Herbicides □
Other Pesticides ▨

Lbs. (Millions)

Source: Ref. *21*.

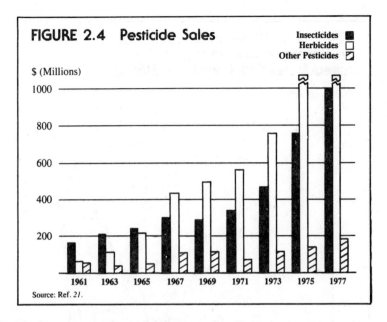

FIGURE 2.4   Pesticide Sales

Insecticides ■
Herbicides □
Other Pesticides ▨

Source: Ref. 21.

BENEFITS   Pesticides have contributed tremendously to improved public health and increased agricultural productivity.   For example, there have been dramatic increases in production of corn, sorghum, soybeans, and wheat per acre from 1930 to 1977 (Figure 2.5).   Increases in yield have made it possible to avoid the proportionately high expenditures on cultivation and harvesting that would be necessary with increased acreages. Chemical control of pests has also helped to prevent inroads on park and wilderness areas.

The benefits of pesticides have been felt throughout the world (especially in the United States), but some of the unintended effects they have produced on the environment have caused increasing concern in the past decade or so.   The use of pesticides should not benefit only consumer, farmer, and manufacturer.   The problem of risk versus benefits for society as a whole must also be evaluated.   Regulatory processes have grown more complex as more data have been accumulated on all aspects of pesticide use.   The analysis of such data, once the sole concern of agriculture, has also become the concern of scientists in various fields and other members of society representing a wide range of disciplines and interests.

Significant crop losses result from direct competition with weeds in the field—for example, losses in Arkansas rice fields ranged from $22 to $105 per acre at 1979 yields and prices (22).   In addition to the direct impact on crops, weeds infest irrigation canals, compete with crop plants for nutrients or water, and interfere with harvesting and grain drying.

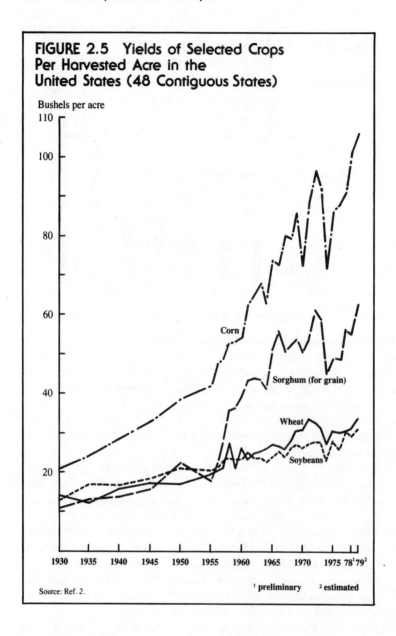

FIGURE 2.5 Yields of Selected Crops Per Harvested Acre in the United States (48 Contiguous States)

Bushels per acre

Corn

Sorghum (for grain)

Wheat

Soybeans

1930 1935 1940 1945 1950 1955 1960 1965 1970 1975 78¹79²

Source: Ref. 2.   ¹ preliminary   ² estimated

Weed control with herbicides results in significant savings in labor and energy. Herbicides that kill and desiccate foliage but are inactivated by soil make it possible to plant crops in untilled soil. Such ''no-till'' systems or similar reduced-tillage conservation practices curtail soil erosion and loss of soil moisture as well as reduce fuel consumption by 50–70%. Each year in the United States, 250 billion tons of soil are moved, much of it several times, in tillage and cultivation operations. This amount of soil would make a ridge 100 feet high and 1 mile wide from New York to San Francisco. At least one-half is practiced solely to control weeds *(23)*.

Weeds and other pests can be controlled to some extent through improved farm management and cultivation and by breeding varieties of crops that are resistant to disease, insects, and many other pests. However, a major factor in pest control has been the use of pesticide chemicals.

LIMITATIONS The growing awareness that increased agricultural production is attained by the intensified use of agricultural chemicals has provoked a variety of responses. Some of the most extreme evoke visions of a totally polluted and poisoned environment. However, dramatic improvements in public health and nutrition since the 1920s and many successful attempts to reduce or eliminate pollution justify a more positive view. Ecological and environmental changes of subtle character do result from pesticide use. A major problem is to define the changes, often on the basis of limited data, and to determine whether they represent adverse or beneficial consequences.

Efficacy. The number of older pesticides in use is continually decreasing. Some may no longer be commercially profitable, while others are no longer effective. Efficacy data are essential to the user and must be provided before a pesticide can be marketed. However, it is important that the need to apply a pesticide be clearly established before use, because routine ''preventive'' applications may have no significant benefit and pose an unnecessary risk to the environment. Another debatable question is that of the benefit of applying pesticides to improve appearance and ''keeping'' qualities. In some cases, such benefits may be considered merely luxuries, although in other cases, blemished food may be unsafe.

Of more practical concern to the farmer is the failure of a pesticide to control pests. Under repeated applications of an insecticide, the process of selection favors the survival of the members of the species that possess the greatest resistance. Ultimately, the resistant species will predominate under continued heavy pesticide use, and another chemical must be substituted. Unfortunately, the insect may have already inherited resistance to new insecticides. Resistant species of weeds are also encountered, but not as frequently as with insects. Nevertheless, the niche formerly occupied by a susceptible weed may be filled by a weed that is resistant to the herbicide formerly used.

Toxicological effects. The long-term toxicological effects of animal exposure to chemicals in general may include mutagenesis, carcinogenesis, or teratogenesis. Emphasis is being placed on the improvement of techniques for acquiring and interpreting toxicological data, particularly data relating to chronic exposure at low dose levels. Improved techniques are urgently needed. As part of this new emphasis, the study of chronic effects of mammalian exposure to low levels of synthetic organic chemicals has intensified. Such studies may not always permit confident extrapolation of safe or unsafe levels of human exposure in the real world; however, efforts to protect those people working directly with pesticides are apparently succeeding. Recent reports indicate that mortality in the United States associated with pesticides has decreased during the past decade, and that most poisonings involved gross mismanagement or carelessness on the part of those handling pesticides *(24)*.

Environmental effects. Some pesticides, particularly metal compounds and some organochlorine insecticides, are extremely resistant to degradation by plants and animals. To avoid frequent applications, a certain degree of persistence or of resistance to breakdown may be desirable in some cases. However, residues of some pesticides accumulate in wildlife, cattle, and humans because they are soluble in fatty tissues and are only slowly metabolized or removed from the tissues. In lactating animals, residues are removed more quickly and end up in the animals' milk. The use of fat-soluble stable pesticides, or disposal of their wastes where they might volatilize or be washed from soil or foliar surfaces into water, can also lead to the accumulation of residues in plants and animals.

Pathways of microbial, animal, and plant metabolism, physicochemical interactions with soil and water, volatility, and photochemical decomposition of pesticides have been studied individually in the field and in the laboratory. Models and miniaturized ecosystems are now used in attempts to simulate these factors and integrate their effects. Such environmental studies are essential for the intelligent use of pesticides, but they contribute substantially to costs.

Integrated approaches. Recognition that, in some situations, adverse effects may accompany continued use of pesticides has led to development of alternative strategies for pest control. These include development of resistant plant species, release of predators and parasites, and use of viruses and pathogens. It is likely that overall approaches that combine many methods to contain pest populations are to be preferred to the exclusive use of a single method.

One advantage of such a combined or integrated approach is that it can overcome the evolutionary adaptation by which a particular target species acquires resistance. Natural regulatory forces in the ecosystem can be maximized to limit the use of conventional pesticides or other measures. Integrated pest control was defined in 1967 by an FAO Panel *(25)* as "a pest management system that utilizes, in the context of the associated environment and the population dynamics of the pest species, all suitable techniques and methods in as compatible a manner as possible and maintains the pest populations at levels below those causing economic injury. . . ." The implication of this definition is that control measures are only used where some economically acceptable injury threshold is exceeded. Successful development of integrated control depends on adequate understanding of the population dynamics of pests and potential pests, of the ecology and economics of the cropping systems, and of possible harmful effects to the general environment *(26)*. Natural forces that regulate pest populations are combined with techniques such as cultural methods; use of resistant crop varieties; release of sterile insects, parasites, predators, or pest-specific diseases; and application of attractants or chemical pesticides.

The term "integrated pest management" has been considered by some authorities as synonymous with "integrated pest control." However, the National Academy of Sciences report on pest control *(27)* pointed out that integrated pest management is broader in that it includes all approaches, from single component techniques to extremely complex control systems. Pest management may utilize techniques such as the establishment of quarantines against introduction of the pest species or the attempted eradication of certain species. All approaches to pest management, including that of not growing a crop in certain areas, may be adopted. The role of natural control practices is maximized, and a variety of techniques are used to enhance the effects of natural forces. Where artificial controls such as pesticides are employed, they may be used judiciously to maintain pest damage potential within acceptable limits.

Integrated pest management is not a revolutionary concept. Instead, it implies only the extrapolation and synthesis of current knowledge from several disciplines for the management of pest problems. If successful, the unnecessary use of pesticides and other artificial control methods should be reduced, but the concept does not imply that their use will cease. Thus, the concept of integrated pest management is based on limiting the use of pesticides to that needed to derive maximum benefits. The term "benefit" may be defined in its widest sense to include the maintenance of environmental integrity, economic crop yields, and public health.

Assessing benefits and risks. Reduction in the use of pesticides to the minimum level consistent with the efficient use of land in crop production will conserve supplies of petroleum. Reduction will prolong the effectiveness of existing pesticides by delaying the development of resistance through adaptation, and there may be also a concomitant reduction in the total cost of pest management.   To achieve these goals, it is important to improve techniques of assessing the benefits and risks of pesticide use in specific situations in the light of available alternatives.   Although social factors are not easily reduced to dollar terms, it is important that they be included in such assessments in both quantitative and objective terms, to provide a sound basis for decision making.

The use of pesticides is changing and the popular concept of the user as a naive countryman has been replaced by that of a technologist familiar with a variety of disciplines.   Pesticides are but a single component of the complex of agricultural chemicals used by the food producer.   While their use will continue to intensify, it falls increasingly under the scrutiny of lawmakers, politicians, and other concerned members of the community.

INTERNATIONAL ASPECTS   A small number of countries—Japan, the United States, and the nations of Western Europe—apply far more pesticides on their cultivated lands than does the remainder of the world (Table 2.2).   However, many other nations require pesticides for crop production, so demand is increasing worldwide.   In 1976 the United States exported 574 million pounds of pesticides; in the same year, 62 million pounds of synthetic organic pesticides and 3 million pounds of natural organic pesticides were imported *(28)*.   Japanese exports find a ready market throughout the Pacific and East Asia; in 1976 the United States imported 5.6 million pounds of pesticides from Japan *(28)*.

In such areas as Asia and Latin America, pesticide production is increasing locally; generally it depends heavily on technology developed in the United States and Europe.   Transfer of such technology may pose a number of problems.   Pesticide production in industrialized countries has been carefully regulated and monitored.   Many pesticides that can be easily and cheaply manufactured—for instance, the organochlorines—may become environmental pollutants or pose severe problems in waste disposal. Moreover, the majority of U.S. pesticides were developed for use in temperate or subtropical zones.   Hence, use patterns must be developed for tropical areas, where the developing nations of the world need to increase food production.

In many former colonial territories where the national economy has been based on monocultures remaining from the colonial era, crop destruction by disease or pests has resulted in widespread distress.   In 1967 and 1968 Arabica coffee production in Kenya was nearly destroyed by disease, but new organic fungicides combined with improved agricultural practices succeeded in saving the crop *(29)*.   In the late 1960s, pesticides controlled

TABLE 2.2   Crop Acreage and Pesticide Expenditures in Different Parts of the World, 1974

Area	Estimated Crop Acres (millions)	Estimated Pesticide Expenditures at User Level (millions of U.S. dollars)	Estimated Pesticide Expenditures Per Acre (U.S. dollars)
Japan	13	450	34.6
Western Europe	128	1,301	10.2
U.S.A.	198	1,732	8.7
All other countries excluding China	1273	1,655	1.3
Total	1612	5,138	

Source: Ref. *18.*

the capsid bug, the most important pest of cocoa in Nigeria and Ghana.   The cost was $6 million, but total benefits were estimated at $66–$130 million *(29).*   The Jamaican banana industry, almost destroyed by Panama disease in 1936, recovered when the Gros Michel variety was replaced by the Lacatan variety *(29).*   However, the latter is susceptible to leaf spot disease and must be sprayed with fungicides to ensure profitable cultivation.

Food production in Asia provides similar examples of the benefits of crop protection.   The development of high yielding varieties of rice and wheat places higher demands on irrigation and fertilizer, which favor an increase of pest populations and the spread of disease.   Insecticides and fungicides are needed to maintain high productivity.   India and Pakistan have substantial crop protection programs, but an FAO symposium recommended that these be expanded from the 50 million acres covered in 1964 to a target of 670 million acres in 1985 *(29).*   Asian countries (except Japan) are extremely dependent on imports of crop protection supplies, and if living standards are to be maintained or improved, the supply of pesticides for food production and public health must be expanded.

FUTURE PROSPECTS   In 1978, the United States was responsible for $2.9 billion, or one-third of the world's pesticide expenditures *(30).*   World demand is projected to grow 14% in the 1980–84 period, while U.S. demand will be up by 17% *(30).*   Production of these materials will require adequate petroleum feedstocks, capital, and energy.   Developing nations should become increasingly involved in production; nevertheless, the United States must continue to play an important role in discovering and developing new pesticides.

Although the shortage of energy and the difficulties of petroleum supplies must be considered in relation to domestic consumption of pesticides, the use of petroleum feedstocks to manufacture pesticides results in a net benefit to the United States.   Pest control represents about 0.2% of the nation's annual energy consumption, while contributing to production of agricultural surpluses that can be exported and so provide dollars to help pay for imported oil.   To help meet world food demands (and to further its own interests), the United States should continue to export both food and pesticides.   As international trade in both pesticides and foods increases, greater international cooperation will be needed in such matters as pesticide registration requirements, pesticide specifications, residue levels, and analytical methodology.

Production Aspects

All of the world's food supplies are derived from plants, either directly or after having been fed to animals and converted to products suitable for human consumption.   Plants grow through chemical reactions and processes. The chemicals necessary for growth are obtained from soil, water, and air through processes in the roots and foliage of plants.

The most important food and feed crops are the cereal grains such as rice, wheat, corn, sorghum, millet, barley, rye, and oats, and the grain legumes and oilseeds such as soybeans, peanuts, peas, beans, and lentils. To meet the world's future food needs, it is estimated that production of cereal grains must double by the beginning of the 21st century, while production of the grain legumes must quadruple *(31)*.   Also important are the root and tuber crops such as potatoes, sweet potatoes, and cassava, and the sugar crops.

Crop production can be increased either by cultivating more land or by improving yields per unit of land.   But suitable land is becoming less and less plentiful, and much of what does remain—including prime agricultural land in many cases—is being given over to homes, highways, and buildings.   Therefore, improved yields hold the key to feeding the world's expanding population, making conservation of prime agricultural land imperative.   While yields are no longer increasing as spectacularly as in the past *(32)*, the potential for further increases remains substantial because a number of major crop yields are well below record yields (Table 2.3) *(33)*.

TABLE 2.3   Average and Record Yields for Selected Commodities

Commodity	Average 1979 [a]	Record [b]	Record Average
Corn (bushel per acre)	106.4	307	2.9
Wheat (bushel per acre)	34.0	216	6.4
Soybeans (bushel per acre)	31.5	110	3.5
Sorghum (bushel per acre)	63.0	320	5.1
Oats (bushel per acre)	53.1	296	5.6
Barley (bushel per acre)	48.9	212	4.3
Sugar beets (ton per acre)	19.9	54	2.7

Source: Ref. 33.
[a] Estimated.
[b] As of 1974.

Meeting future food goals will require more effective use of land, water, energy, manpower, machinery, fertilizers, and pesticides. But in addition, new technology should be developed from fundamental research directed toward basic physiological and biochemical processes, rather than just toward individual crops as it has been in the past. Hardy *(34)* has called the limitations revealed by fundamental research as the "what's wrong" approach, and suggests coupling chemistry with biology to identify and correct specific limitations inherent in the plant. Evolution results in enhanced ability for the survival of a species, not necessarily the maximum production of economic crops for man. Therefore, insofar as crop production is concerned, numerous improvements could be made in the plants themselves and in the processes by which they produce food.

GENETIC IMPROVEMENTS   Breeding by conventional techniques, including hybridization, is coupled with labor-intensive selection of improved plants. These techniques have been successful in the past in increasing the yields of various crops and can be expected to continue to do so in some crops. A key feature of the "Green Revolution" was the development of new varieties of higher yielding crops, especially wheat and rice. The conventional techniques, however, are slow—a growing season is required for the experimental plant to mature, many crosses and selections are required to find improved plants, and selections are usually made using only easily observable characteristics. Furthermore, crosses are almost exclusively limited to crops of the same genus, such as wheat crossed with wheat (potentially valuable crosses between genera, such as wheat and soybean, are precluded).

In the future, genetics is likely to depend heavily on the molecular approach, using tools such as mutation breeding, cell culture, fusion of vegetative cells (cells other than reproductive cells), and recombinant DNA *(35)*. These tools have many potential advantages. They may be able to speed up the process of developing improved crop varieties. They may also help increase the genetic diversity of plants. Such diversity provides plant breeders with a pool on which to draw, as they try to develop new varieties with desirable characteristics that result in higher yields and resistance to such environmental stresses as pests, air pollutants, temperature extremes, drought, or saline water.

Treating seeds to change the genes, and hence the plant that grows from them, can increase genetic diversity. The technique, called mutation breeding, can use either physical means, such as x-rays or ultraviolet light, or chemicals. Among the new varieties created by mutation breeding are sorghum, corn, and barley with improved protein quality. The new varieties have smaller amounts of prolamines, a protein of low nutritional value, and larger amounts of lysine, an essential amino acid of high nutritional value.

Improving the quality of cereal protein is important because it is the principal source of protein in the developing countries and among low income groups. With the cost of animal protein steadily increasing, plants will have to provide a greater share of protein intake. Because most cereal proteins are deficient in certain amino acids, grains and legumes must be combined in the diet to provide adequate amino acid balance. Improvement of cereal grain protein through mutation breeding offers the possibility of providing adequate balance while minimizing the need for legumes or animal proteins.

Despite some successes, mutation breeding has not lived up to earlier expectations, because the processes that take place in mutated cells are not well understood. According to a National Research Council study *(36)*, mutation breeding deserves serious reexamination in the light of the need to increase crop yields.

Cell culture is a major technological breakthrough of the early 1970s. It may greatly reduce the time required—and provide opportunities previously unavailable—to manipulate plants, and it may also enable transfer of genes between genera *(37)*. Large, relatively homogeneous populations of vegetative cells can be grown in test tubes. In a few cases, among them corn, these cells can be made to grow into whole plants, while for other plants, including soybeans, this essential step has not yet been achieved. In several cases, the first steps to hybridization of vegetative cells have been achieved by, for example, fusing a soybean cell with that of a cereal grain such as wheat. Chemicals, among them sodium nitrate and polyethylene glycol, are used to cause the fusion.   Unconventional breeding of plants without recourse to the normal sexual reproduction is not yet a practical tool for increasing crop yields; it is, however, a promising shortcut that may one day be a valuable adjunct to conventional techniques.

Application of the techniques of recombinant DNA to improving plant yields is at an even earlier stage of development.   The potent and much publicized technique, which has already been applied to production of human insulin, involves the removal of the desired DNA from one organism, its attachment to a carrier DNA, and insertion of this recombined DNA into the cell of the plant whose genetics are to be modified.   An example might be the incorporation of the genes for nitrogen fixation into corn.   A carrier system that is under intense study is *Agrobacterium tumefaciens,* a bacterium that can infect a wide range of plants, producing a gall, or tumor *(37).*   The recombinant DNA technique will enable for the first time the application of chemical and biochemical techniques for a directed, rather than random, manipulation of plant genetics.

PHOTOSYNTHESIS   Foremost among the biological processes that must be improved if crop yields are to be increased is photosynthesis, the primary process for all food production *(38, 39).*   Using electromagnetic energy in the visible wavelengths of sunlight, photosynthesis converts the carbon dioxide in air to sugars and starches (plus oxygen) and then to other plant constituents.   Photosynthesis is responsible for producing up to 95% of the dry weight of plants *(40).*

Despite its importance, photosynthesis research in the United States was funded with only $10 million in 1976 *(41).*   Most food crops convert less than 1% of the sunlight reaching the plant's leaves into food, making this essential process a promising research target.   Crop production is determined by the gross uptake of carbon dioxide during photosynthesis minus the losses of carbon dioxide during respiration, which occurs in all green plants and algae.   Some respiration is needed to provide the energy necessary for growth and maintenance.   There is some evidence that not all this respiration is essential; thus, if part of it could be eliminated, crop yields would probably increase.

In the past quarter century, scientists have discovered a second and apparently wasteful form of respiration, which is known as photorespiration because it occurs at a high rate in light in most plants.   In plants having a high rate of photorespiration, there is one biochemical cycle for photoassimilating carbon dioxide.   Plants with this one cycle are called C_3 plants, because the first compounds formed when carbon dioxide is assimilated are acids containing three carbon atoms.   In some plants, including sugarcane, sorghum, and corn, there is an additional carbon dioxide fixation cycle which leads to four-carbon acids. C_4 plants have a decided advantage because they carry out little photorespiration, which results in the recycling of as much as 50% of the carbon dioxide that the plant had fixed. In general, C_4 plants have higher average yields than C_3 plants, according to figures from the U.S. Department of Agriculture. In 1979, the average annual yields per acre were 102 bushels of corn and 60 bushels of sorghum. For two C_3 plants, the yields were 34 bushels of wheat and 30 bushels of soybeans.

Scientists are taking various approaches to reducing photorespiration. One is through the use of chemicals. Another is through the attempt to endow C_3 plants with C_4 characteristics (42). The opportunities in this genetic approach are great, and research activity is intense. While some successful results have been reported, other scientists have not been able to reproduce those results.

The initial enzyme required for the biochemical conversion of carbon dioxide to carbohydrate in photosynthesis is ribulose-1,5-bisphosphate carboxylase. This enzyme is the most abundant in nature, constituting up to 50% of the soluble protein in plant leaves. However, it reacts not only with carbon dioxide, but also with oxygen as part of the photorespiratory process. Understanding and controlling these competing reactions could lead to greater assimilation of carbon dioxide and hence, more efficient photosynthesis.

The competition between oxygen and carbon dioxide in the ribulose-1,5-bisphosphate carboxylase reaction results, at least in part, from the high concentration of oxygen in the atmosphere (21%) and the low concentration of carbon dioxide (0.03%). The level of carbon dioxide is sometimes increased in greenhouses, which increases the dry weight of tomatoes, lettuce, and other crops by 50–100%. In field experiments in which air enriched with carbon dioxide was added in large amounts near the soil within open-top enclosures, yields of grain legumes, including soybeans, peanuts, peas, and beans, increased by 50% or more (34). In an open field, unrealistically large amounts of carbon dioxide would be needed because it would diffuse quickly into the atmosphere. Hence, this technique of overcoming photorespiration will not be practical for field-grown crops. Nevertheless, the experiment demonstrates that yields can be greatly increased when photorespiration is decreased. Aside from photorespiration, other opportunities for increasing photosynthesis will undoubtedly be identified as the complex process is better understood.

NITROGEN FIXATION   Nitrogen, which makes up 78% of the earth's atmosphere, occurs in every protein molecule and many other molecules that form living tissues. However, with few exceptions, plants and animals can assimilate the nitrogen they need only when it is combined, or fixed, with other elements. Therefore, to supplement the nitrogen found in soil, nitrogen fertilizer is frequently applied to crops. Nitrogen fertilization has been one of the major contributors to the doubling of world grain production in the past quarter century.

Meeting crop production goals for the beginning of the 21st century will require quadrupling nitrogen fertilizer production. This would require construction of an additional 500 nitrogen plants, each costing at least $100 million at 1979 prices *(43)*. The problems of such expansion, including the cost and the availability of fossil fuels, have focused attention on other alternatives. One is to develop new catalysts that would permit production of ammonia from nitrogen in the atmosphere at lower temperatures and pressures than the present process utilizes, thereby reducing the cost both in dollars and energy. This new phase of nitrogen fixation chemistry was launched in the early 1960s when Soviet, Canadian, British, and American scientists reported a number of different chemical systems involving transition metal catalysts that, in the laboratory, can produce ammonia under mild conditions.

An effective method of using such an ammonia producing system would be to link it directly to a crop in the field; for example, the system could be tied in with the irrigation stream and synchronized so as to match the changing nitrogen needs of the crop throughout its growth cycle. Such an arrangement would also eliminate the transportation, storage, and application costs of conventional methods of nitrogen fertilization. Each step required for such a system can be performed in the laboratory. However, the catalyst must be improved greatly before the system could be of practical significance. Thus, while large-scale use of better nitrogen fixation systems appears to be decades away, the stakes in their development are large, and further research may well develop a technology that could help meet the world's increasing nitrogen requirements.

Another possible solution to meeting nitrogen requirements lies in the ability of certain bacteria and blue-green algae (that live in fields, forests, and oceans) to fix nitrogen. These organisms possess the genetic information to synthesize an enzyme, nitrogenase, that catalyzes the conversion of nitrogen to ammonia at normal temperatures and pressures. Worldwide, an estimated 265 million metric tons of nitrogen are fixed annually, of which about 175 million are fixed biologically *(44)*. About 90 million metric tons are fixed in agricultural areas, about equally divided between croplands and permanent meadows (Table 2.4). A major part of the nitrogen fixed in agricultural soils is supplied by bacteria that establish a symbiotic association with legumes. Since biblical times, legumes have been plowed back into the soil as a "green manure" to increase its nitrogen content. However, spending an entire crop for nitrogen fertilization has proved uneconomical, and crop rotation is used only in developing countries.

TABLE 2.4   Estimates of Nitrogen Fixation; Past, Present, and Future (in Millions of Metric Tons)

| | | | 2000 | |
| | | | Expansion of Current Technology | Introduction of New Technology |
Type of fixation	1900	1977		
Biological	155	175	195	265
Agricultural		90		
Forest and woodland		40		
Unused land		10		
Sea		35		
Nonbiological		90	210–260	70–80
Industrial		60	180–220	40
Combustion and lighting	15	30	30–40	30–40
TOTAL	170	265	405–455	330–340

Source: Ref. *31*.

In the symbiosis between a legume and *Rhizobium* bacteria, specialized nodules are formed on the legume roots.   Through their normal metabolic processes, the bacteria provide nitrogen-containing products to the host plant, making it at least partially self-sufficient in nitrogen; in return, the bacteria receive nutrients and protection from the host.   However, the symbiosis is highly specific, so that in general each legume requires a different *Rhizobium* species.   Since early in the 20th century, *Rhizobium* has been sold commercially for inoculation of legumes.   However, the inoculation technology fails to assure that an applied *Rhizobium*, though superior, can compete successfully with the less desirable *Rhizobium* strains already present in most agricultural spoils.   A beneficial response is often difficult to demonstrate in fields where the crop has been grown previously. A major need is improvement of the efficiency of the *Rhizobium*-legume systems *(45)*.

In another symbiotic partnership, a blue-green algae, *Anabaena*, lives in leaf cavities of a water fern *(46)*.   This partnership is exploited in the Orient by encouraging the growth of the fern in flooded rice paddies to provide a natural source of fixed nitrogen.   This attractive system is being unravelled in recent fundamental studies, while agronomic approaches seek to improve its utility.   Nitrogen can also be fixed by other bacteria and blue-green algae that are free-living entities.

Despite the diversification and varied associations, all nitrogen-fixing organisms contain just a single biological catalyst: the enzyme nitrogenase.

Its existence was postulated in 1934, but it was not extracted from a bacterium until 1960 *(47)*. Since then, a great deal has been learned about the enzyme, although its exact composition and how it catalyzes nitrogen fixation are still unknown. However, enough is known to say that, like the photosynthesis enzyme, there is much that is wrong with it in terms of economic cropping *(31)*. Nitrogenase is a large molecule. It is "poisoned" irreversibly by oxygen, unstable at both high and low temperatures, and sluggish in reactions. In addition, it consumes great amounts of energy—25 molecules of adenosine triphosphate (the basic energy currency of all living systems) per molecule of nitrogen fixed. If chemical and biological researchers can learn to modify or control the enzyme's activity, they may be able to increase nitrogen fixation in legumes as well as extend nitrogen fixation to important grains such as corn and wheat, and to identify new and cheaper pathways for nonbiological fixation.

Several other recent research findings encourage the prospects for enhancing nitrogen fixation in legumes. In legume experiments on carbon dioxide enrichment, nitrogen fixation was increased several-fold as a result of increased photosynthesis *(48)*. The increase in fixation of nitrogen arose from an immediate increase in nitrogenase activity, a long-term increase in nodule mass and nitrogenase activity, and a prolongation of fixing activity. These experiments clearly establish photosynthesis as the principal factor governing nitrogen fixation in legumes, and probably in all biological nitrogen-fixing technologies.

Almost simultaneous reports in 1975 from five laboratories in Australia, Canada, and the United States indicate that *Rhizobium* can be made to fix nitrogen independently of an association with legumes, thus raising the possibility that a promiscuous *Rhizobium* could be developed that would provide nitrogen to nonlegume crops. This revolutionary advance has nullified the 75-year-old belief that *Rhizobium* can fix nitrogen only with legumes and proved that *Rhizobium* is indeed the source of the genes for the legume–*Rhizobium* association *(49)*. The belief that the associations are exclusive was also challenged by the chance discovery by an Australian scientist of a *Rhizobium* nodular association in a nonlegume shrub. These discoveries suggest that nature possesses much more undiscovered diversity, some of which may be of great value in development of new nitrogen-fixing technologies.

Rather than enhancing nitrogen fixation in legumes or extending it to cereal grains, transferring the genetic information for fixation directly to plants is the ideal solution, because it would eliminate the complexities of putting together two biological systems, the bacterium and the plant. Molecular biological studies of the genetics of nitrogen fixation started in the early 1970s, when American scientists identified several genes responsible for nitrogen fixation in bacteria.   The number has now increased to about 14, and recombinant DNA techniques have been used to construct hybrid DNA for the transfer of some of these genes to other bacteria and possibly eventually even to plant cells.   Meanwhile, U.S. expenditures on nitrogen fixation research are modest, totalling $5 million in 1976 *(41)*.   There are, however, many promising avenues of research that, with adequate support, could lead to practical technologies to improve nitrogen fixation in both biological and nonbiological systems.

PLANT GROWTH REGULATORS   Plant growth regulators are natural or synthetic organic chemicals that, when applied directly to a plant, change its life processes or structure in some beneficial way *(50)*.   Initially, regulation of plant growth was considered the exclusive domain of certain naturally occurring plant hormones.   Now, recognizing that all molecules and processes involved in growth and development are subject to manipulation, scientists are studying such key factors as enzymes (including ribulose-1,5-bisphosphate carboxylase), and movement of major molecules to the harvested part of the plant *(51)*.   In such crops as soybeans and corn, a plant growth regulator would seek to maximize the growth of the seed, while in others—for example, alfalfa and potatoes—maximum growth should be in roots, stems, or leaves.

The most promising goal of plant growth regulation is to enhance yield.   The range of opportunities is broad, extending from regulation of seed germination and root growth to control of senescence (aging) of the mature plant.   Plant growth regulators can also be used to improve the nutritional quality of plants (for example, the protein or oil content) and their aesthetic characteristics (for example, uniform size and color).   In addition, plant growth regulators can facilitate production.   For instance, they can induce uniform ripening and so permit mechanical instead of hand-picking.

Plant growth regulators have been used since the 1930s, when ethylene was applied to induce flowering of pineapples (Table 2.5).   A number have been introduced since then (see Chapter 3 for a discussion of their use in fruits and vegetables).   With a few exceptions, however, none of the plant growth regulators have been for use in major agronomic crops.   As a result, sales are small: in 1980, they are expected to represent only 1% of the total world agrichemical market of $8 billion *(51)*.   Several factors are responsible for this slow development, including an empirical rather than fundamental approach and the emphasis on hormones and horticultural crops. These factors are now changing.   In addition, 30 companies are reported to be working on development of plant growth regulators.   A Plant Growth Regulator Working Group was organized in the United States in 1973, and a similar professional organization was recently set up in Europe. All these activities, coupled with an expanding understanding of the chemistry of plants, may lead to new plant growth regulators that could one day make as big a contribution to increased crop production as fertilizers, pesticides, and improved plant varieties have in the past *(50, 51)*.

TABLE 2.5   Selected Plant Growth Regulators Introduced Since The 1930s

Time	Growth Regulator	Use
1932	Ethylene	Pineapple flowering
1940s	Naphthalene acetic acid	Setting and dropping of fruit
1950s	Maleic hydrazide	Inhibitor of potato and onion sprouting
	Gibberellin	Delay ripening
1960s	Chlormequat chloride	Wheat dwarfing
	2,3,5-Triiodobenzoic acid	Soybean dwarfing (discontinued)
	Daminozide	Control of size and color of various fruits and peanuts
1970s	Ethephon	Ripening of various fruits and enhancing of color
	Glyphosine	Sugarcane ripener
	Cycloheximide	Citrus harvest aid
	2-Chloro-6-trichloro-methylpyridine	Nitrification inhibitor
	Etacelasil	Olive harvest aid
	5-Chloro-3-methyl-4-nitro-N-pyrazole	Citrus harvest aid
	Glyoxal dioxime or ethane-diol dioxime	Citrus harvest aid

Source: Ref. *51*.

ANIMAL PRODUCTION

Animal products contribute more than half of the total nutrients of the average diet in the United States, including two-thirds of the protein, one-half of the fat, one-third of the calories, four-fifths of the calcium, and two-thirds of the phosphorus. Globally, 20–25% of man's protein requirements and 10% of the calories are provided by livestock and poultry *(52)*.

An estimated two-thirds of the feed used in producing the world's animal foods comes from substances that are undesirable or unusable for human consumption. About two-thirds of the world's agricultural land is

in permanent pastures, rangeland, and meadows, of which 60% is not suitable for cultivation, but can be used for grazing purposes.   Furthermore, on
most cropland, almost half the total digestible energy produced by the plant
is left on the fields after harvest *(53)*.   Ruminants (cattle, sheep, and goats)
serve as "protein factories," converting plant residues unfit for human
consumption into edible foods of excellent quality and high acceptability.
Animals, both ruminants and nonruminants, also utilize huge tonnages of
by-products and wastes from food and industrial processing, materials that
might otherwise pollute the environment.   The remainder of the materials
fed to animals is largely cereal grains.

General Outlook for Animal Production

It appears that at least to the year 2000, the animal industries in the United
States can continue not only to provide the animal products needed domes-
tically, but also to provide surpluses for export.   If grains, soybeans, and
other edible feeds are needed in higher quantities for direct human consumption, animals can be fed higher levels of forages (grass, hay, crop
residues), plus other materials not consumed by humans.

The developing countries have considerable potential, as well as major
constraints, for increasing animal production, primarily through greater efficiency in feeding and management practices and through improved breeds
of livestock.   Presently, developing countries have 60% of the world's
livestock and poultry but produce only 22% of the world's meat, milk, and
eggs.   By contrast, the United States, with only 8% of the world's animal
population, produces 19% of the world's meat, milk, and eggs
(54).   Many developing countries are limited in the land and other resources required for expansion; overgrazing and desertification are widespread problems.   But even in such countries, the limited amounts of additional animal protein foods that could be produced without undue strain on
the environment and cereal grain supplies would be valuable in supple-
menting the basic diet.

Feeding and Nutrition of Land Animals

During the past 50 years, developments in chemistry and biochemistry have
resulted in great progress in the science of feeding and nutrition of animals.
Once fed almost entirely on feeds available in fields or barnyards, many
animals now are at least partially fed on scientifically blended rations designed to promote good nutrition and health in an efficient manner—the
highly organized production of poultry being a good example.

VITAMINS   In the early 1900s, forages provided the vitamins for animal feeding.   Beginning in the 1920s and accelerating in the following decades, animals have grown faster and produced more, as animal feeding, selection, and management have become more sophisticated.   As intensified and confined production operations evolved with greater emphasis on formulation of diets, it then became necessary to take several measures to ensure the adequacy of various vitamins in animal diets.   Materials such as tankage, fish meal, milk, and forage were used for this purpose.   In addition, the availability of reasonably priced concentrates or purified forms of vitamins was an essential part of this development.   Today, most ruminant diets are supplemented with vitamins A and D and sometimes with vitamin E.   Most swine and poultry diets are now supplemented with vitamins A, D, E, K, riboflavin, niacin, pantothenic acid, and choline; under certain conditions, biotin and B_6 are also used.   An especially significant development was the identification in the 1940s of vitamin B_{12}, the "animal protein factor," which enabled the use of swine and poultry rations derived exclusively from plants.

MINERALS   For some decades, 13 mineral elements (sodium, chlorine, calcium, phosphorus, sulfur, potassium, magnesium, manganese, iron, copper, cobalt, iodine, and zinc) have been recognized as performing essential functions in an animal's body.   Evidence is increasing to show that selenium, molybdenum, and fluorine also perform essential functions.   There is some indication that nickel, vanadium, chromium, tin, silicon, cadmium, and arsenic may also have essential roles.   Other factors—including new methods of processing feeds, higher yields per acre of forages and crops, new plant varieties, new fertilizer formulations, processing methods, higher-producing animals, and declining soil fertility—may increase the need to add other mineral elements to animal diets.

AMINO ACIDS   When animals are fed natural products in their diets, they frequently do not receive adequate amounts of certain amino acids. Thus, amino acids such as methionine and lysine are now being added to many poultry and swine diets to provide optimum amino acid balance. Combinations of various feed protein sources also can provide favorable amino acid balance.

NONPROTEIN NITROGEN   Alone among animals, ruminants are able to convert simple nitrogen compounds (via microorganisms in the rumen) into a substantial proportion of the protein they require, thus reducing their need to consume edible materials.   Urea ($CO(NH_2)_2$) is the most widely used source of nonprotein nitrogen (NPN) *(55)*.   According to estimates of the U.S. Department of Agriculture, 468,000 tons of urea were used in 1978 in ruminant diets in the United States, taking the place of 1.3 million tons of protein that would otherwise have been used.   NPN is most efficiently used in diets rich in soluble carbohydrates such as molasses and is not efficiently used in low-quality forage diets.

The challenge in the future is to make NPN more effective and safe in low-quality forage diets.   Most scientists now recommend that urea can safely supply at least one-third of the total protein in ruminant diets.   Proper management of diet mixing and feeding, however, is necessary when urea is used at such high levels.   Well-controlled experiments have shown that it may be possible to provide as much as 50% of the protein requirements of ruminants through NPN.   Biuret ($(CO)_2(NH_2)_2NH$), which is produced by heating urea, has been used successfully to supply nitrogen for ruminants being fed low-quality forages.   It slowly reacts with water to produce urea, so that it is less toxic than urea.   However, it is considerably more expensive.   Thus chemists could make a major contribution by providing new effective, safe, and inexpensive sources of NPN.

Nonnutritive Feed Additives

Since the early 1950s, a number of drugs have been added to animal feeds to control disease, promote growth, and improve feed efficiency.   Among these nonnutritive additives are hormones, antibiotics, sulfonamides, nitrofurans, and arsenicals.   A number of feed additives have had a major effect in increasing the efficiency of animal production.   For example, by stimulating the growth of cattle, the hormone diethylstilbestrol (DES) can reduce by 10–12% the amount of feed required, for a saving of 7.7 billion pounds of cattle feed annually in the United States *(56)*.   Discontinuing the use of DES would increase retail expenditures for beef by about $480 million per year *(56)*.

The safety of DES has been questioned in recent years because any residues that remain in edible tissues of animals may cause cancer in humans.   Hence, in 1972, the FDA withdrew approval of DES in livestock feed.   The U.S. Court of Appeals later vacated the order.   FDA began new proceedings and finally, in 1979, banned the use of DES in livestock feed. There are other drugs with FDA approval that can replace DES.

The antibiotics added to feed at low levels increase the weight gain of animals, primarily by preventing disease.   On the basis of the 1975 average yield in the United States, about 2 million acres of land would be required to produce the feed saved by the use of antibiotics with cattle and swine *(57)*.   It has been estimated that the use of antibiotics with cattle, calves, and swine saved the consumer $2 billion in 1973 *(58)*.   Antibiotics are also of value for poultry and other animals, but no estimate has been made of dollar savings.

In 1977, once more due to potential risk to human health, the FDA announced plans to prohibit the use of penicillin and restrict the use of tetracyclines.   The potential risk of feeding low levels of antibiotics to animals lies in creating resistance in bacteria to medications often used to treat human illnesses.   Physicians are now reporting reduced effectiveness of these same antibiotics in treating human disease, according to a study of antibacterial agents and DES made by the Office of Technology Assessment (OTA) *(59)*.   The FDA proposal is pending.   Meanwhile, the U.S. Congress appropriated funds for the National Academy of Sciences to conduct a study on antibacterial agents used in animal feed.

The use of feed additives over the past 25 years has been a major factor in revolutionizing the raising of livestock and poultry.   The result has been larger supplies of meat, which have helped moderate inflationary rises in food prices.   But the immediate economic benefits may be accompanied by long-term risks to human health that have not been evaluated.   The OTA report points out that there is much disagreement among scientists of the validity of many of the findings and the weight that should be attributed to them when considering restricting use of drugs in livestock feed.

Reproduction and Genetics

Many alternatives exist for improving animal production by improving fertility *(41,61,62)*.   World ruminant reproduction rates, for example, are less than two-thirds of potential.   The prostaglandins, a family of hormones used to treat human reproductive problems, are being studied as a means of controlling the reproductive cycle and greatly facilitating artificial insemination.

Selective breeding, once directed primarily towards purebred show animals, is now being directed toward improving disease resistance, efficiency of feed conversion, growth rate, carcass or product quality, and reproductive efficiency.   The poultry industry has demonstrated the improvements that can be achieved.   Production of hybrid poultry began to increase in the United States in the 1930s and, along with advances in nutrition and disease control, led to an increase in the average weight of chickens from 1.7 pounds in 1930 to 2.6 in 1956, while the feed required per pound produced was halved *(60)*.

Intensified Animal Production and Infectious Diseases

Losses from infectious diseases cut animal production in the United States by 15–20% yearly.   In developing countries, where veterinary medicine is not as advanced, losses ranging between 30–40% are common.   As intensified animal production increases, certain infectious disease problems might increase due to close confinement of animals, which exposes them to greater numbers of the infectious organisms; in any case, if a disease does strike, it is likely to affect a large number of animals in a single episode. Thus, animals in large units are more vulnerable to disease than animals in small ones. On the other hand, parasite problems may tend to be reduced by intensive confinement operations, because there is less chance for the organisms to complete their life cycle in certain kinds of units, such as those that have slotted floors or other arrangements that minimize contact with feces.

A recent case demonstrates the potential of research on animal disease prevention. A U.S. vaccine against Marek's disease in poultry was used worldwide following its development in 1969 *(52)*. In the United States, it reduced losses in broilers from 1.57% in 1970 to less than 0.25% in 1974. Over the same period, the mortality rate of laying hens was reduced by 14.8%. The return on the research investment for this vaccine has been estimated at 220% per year.

Future Outlook for Animal Production

A number of recommendations for research to increase animal production have been made by the National Research Council's "World Food and Nutrition Study," *(52)* as well as by other studies *(60, 61)*. Chemistry and biochemistry will be in the forefront of these new developments, which may include:

• *Increasing food output per animal unit* by developing alternate methods of growth stimulations by using natural hormones or approved synthetic compounds; influencing the animals' digestive microbial population; and reducing the use of potentially hazardous materials.

• *Increasing feed resources* by improving existing feed sources through the development of better processing methods, the elimination of natural inhibitors, the reduction of the presence of harmful substances, and the development of grain and forage species of higher nutritive value; substituting food and industrial by-products for grains; developing new nonprotein nitrogen sources for ruminants; increasing the use of single-cell and nonconventional proteins as sources of feed; and utilizing animal wastes as feed.

• *Reducing disease losses* by eliminating the major epidemic infectious disease; controlling endemic disease; controlling parasites; increasing animal resistance to infections via basic immunology; and developing more effective and economical chemotherapeutic agents.

FISH PRODUCTION

Aquaculture—the cultivation and harvest of both freshwater and marine aquatic species—has developed more slowly in the United States than have other sources of food *(62)*.   Of the 6 million metric tons of fish and shell-fish produced worldwide in 1975, the United States accounted for only 65,000 metric tons.   Among developed nations, Japan, for example, produced 500,000 metric tons in 1975, up from 100,000 metric tons in 1971. Many protein-short developing nations in the tropic and subtropic regions of the world have great potential for aquaculture, because water temperatures are favorable for rapid growth year round and labor costs are low.

Because of the confinement of aquatic species to water, the problems posed are different from those involved in raising terrestrial animals, which are affected directly by land and air as well as water *(62)*.   One problem is that the feeds must perform well in water.   Another is that fish are especially sensitive to certain substances such as pesticidal and other organochlorine compounds and heavy metals.   On the other hand, fish readily metabolize the nucleic acids in the single-cell protein (SCP) produced by yeast and bacteria.   The presence of nucleic acids limits the use of SCP in human foods *(63)*.   Fish are also efficient converters of feed into proteins *(64)*.   For example, trout produce almost 20 times as much protein from the calories they consume as do beef cattle (Table 2.6).

TABLE 2.6   Grams of Protein Produced per Megacalorie of Food Ingested by Selected Animals

Animal	Grams of Protein Produced per Megacalorie of Food Ingested
Beef cattle	2
Swine	6–8
Poultry broilers	10–12
Dairy cows (total production)	15–20
Trout	35–40

Source: Ref. *64*.

Early Use of Chemicals

Chemicals and the principles of industrial chemistry have long been used in the production of fish as well as in their preservation and processing *(65)*.   A manual published in 1900 describes several treatments for external parasites—dipping fish in a mild solution of formaldehyde for a few moments, placing them in a 3–5% rocksalt solution until they showed signs of stress, or using mild concentrations of acetic acid *(66)*.

The first diets, which consisted of wet mixutures of slaughterhouse and fishery by-products, were often treated with salt to improve their consistency.   Fish cultured in ponds depended upon fertilization of the waters with compounds containing nitrogen, phosphorus, and potash.   Between harvests, ponds were drained, and the soil was treated with lime to recondition the surface and to attempt to kill pathogens.

Later, development of intensive fish husbandry techniques for production of fish for food or for replacement stocks in ponds, streams, and impoundments has been accompanied by an increase in the use of a variety of chemicals to control the environment, nutrition, health, and population levels of concentrated fish populations *(67)*.

Preparation of Modern Fish Feeds

Supplemental and complete diets are produced on a large scale for fish and crustacea reared by modern fish production methods *(67)*.   Most fish diets are manufactured by grinding, sieving, and mixing; pelleting, extruding or compressing; and drying, freezing, or crumbling.   These diets require most of the traditional nutrients as well as ingredients peculiar to the needs of aquaculture.

PROTEINS   Protein supplements are very important in formulation of fish diets.   Because most fish reared in the Western Hemisphere sell at prices comparable to meat, the fish farmer can afford to pay premium prices for feed.   However, in the future, many traditional fish feed ingredients may be used as human food, forcing the fish industry to find new feed materials not suitable for terrestrial farm animals or humans (62).

Oilseed meals have consistently provided a portion of the protein component in diets.   In addition, large amounts of products derived from processing of slaughterhouse and fishery wastes—including fish protein concentrates, blood meal, meat and bone meal, poultry by-product meal, and hydrolyzed feather meal—are incorporated into commercial fish feeds. Other potential protein sources are SCP, yeasts, and bacteria grown on agricultural or industrial waste materials or derived from the fermentation industries, plankton, and algae.

Producing proteins from algae requires monitoring and controlling the nutrients and pH of the water to assure the maximum yield of the desired species. Direct harvesting of the algae generally is uneconomical at present, but in many tropical areas of the world, the algae produced under abundant ultraviolet radiation in ponds, impoundments, and estuaries are subsequently cropped by various species of herbivorous fish *(63)*. New up-welling techniques have proved successful in raising nutrient-rich water from the depths, which can then be used to produce heavy concentrations of algae near the surface.

VITAMINS   For many years, vitamin deficiencies in fish were suspected on the basis of fragmentary observations. Specific deficiencies were not fully confirmed, however, until 1953 when adequate test diets were formulated using purified vitamins. Shortly thereafter, the vitamin requirements for trout, catfish, and carp were published, together with descriptions of the specific deficiency syndromes *(67)*. Eleven water-soluble and three fat-soluble vitamins have been found to be essential for most fish *(68)*. As a result, these micronutrients are included routinely in most commercial diet formulations. Laboratories throughout the world are determining specific vitamin requirements for other species of fish reared locally.

New intermediates or derivatives of vitamins can also help improve fish diets. Converting heat-sensitive vitamins to heat-resistant compounds such as ascorbic-2 sulfate will extend the shelf-life of the finished feed and in addition ensure the vitamin content in a formulation *(69)*. Various salts of other vitamins may be more available to the fish or prevent rapid leaching of the vitamin into the water.

MINERALS   Fish food rations require a number of minerals. Although fish can generally obtain at least some of the calcium they need from the water supply, additional phosphorus, magnesium, iron, sodium, and potassium must be added to the diet as mixtures of various chemical salts *(70, 71)*. Research is underway to define the specific requirements of young, growing, and brood-stock fishes. As these requirements are defined, the chemical industry will be able to provide the mineral preparations having the desired characteristics.

AMINO ACIDS   Many of the fish diets formulated from agricultural commodities lack adequate amounts of certain amino acids, including lysine, methionine, arginine, and tryptophan *(63)*. SCP supplements are often deficient in methionine, while cereal-grain protein are generally deficient in lysine. Supplementation with synthetic amino acids may allow the use of lower cost commodities for formulation of fish diets, depending on economic considerations.

FATS   Fish, like other animals, require certain unsaturated fatty acids *(72)*. Oils chemically produced from menhaden, herring, salmon, and cod have been used in fish diets for many years. Appropriately processed vegetable oils such as cottonseed, soybean, linseed, sunflower seed, and safflower seed are alternate sources for fish food. Pollack residual oil, squid oil, and poultry by-products oil also can be used in fish diets provided any deleterious compounds are removed.

FEED "BINDERS"   The first "wet" fish diets consisted of minced tissue of animal by-products and fish, plus some plant products. Mixed diets containing both wet and dry ingredients were developed later. Today, dry compressed pellets of ground ingredients are the most common type. Extruded diets (usually frozen) and agglomerates of finely ground ingredients are also used.

One or more chemical "binders" are often incorporated to achieve the desired characteristics in the final product. In wet feed mixtures, salt and agar or some carbohydrate derivative such as carboxymethylcellulose have been used to lend a rubbery consistency to the minced diet. Frozen diets consisting of both wet and dry ingredients often include cellulose derivatives or industrial starches to achieve a more tightly bound material before freezing. Rolled or agglomerate diets incorporate industrial starch, industrial gelatin, or agar to bind the fine particles. Other types of diets may include agar or commercial gums to yield desired consistency. Many feeds are sprayed with a digestible oil to coat the pellets and to enhance stability in water *(73)*.

ANTIOXIDANTS   Antioxidants play an important role in the manufacturing, processing, storage, and use of fish feeds because the fatty acids and some of the vitamins required in the diet are prone to oxidation *(67, 70, 71)*. Conventional fish meal contains residues that can rapidly oxidize to form toxic and growth-inhibiting materials. To prevent oxidation, some fish meal and finished fish feeds incorporate materials such as butylated hydroxyanisole (BHA), butylated hydroxytoluene (BHT), or ethoxyquin. The tocopherols present in most vegetable oils help prevent oxidation, and purified tocopherol is often included in the vitamin supplement or oil mixture used in formulating feeds. These antioxidants are also used as food additives. More effective antioxidants are needed to increase the shelf-life of fish feeds.

GROWTH PROMOTERS   Growth promoters or stimulants have been considered for incorporation in fish feeds. Iodinated casein has been used to assure maximum content of thyroid hormone in young, growing fish. Diethylstilbestrol and growth hormones also have been studied for use as growth promoters.

ATTRACTANTS   Since some fish species need to be taught to feed on artificial diets, attractants have been incorporated into fish feeds to enhance acceptability.   Several amines have been found to be attractive to fin fish and shellfish.   Some amino acids have been tested in salmon, eels, carp, and prawns.   Bait that incorporates attractants has been marketed for crabs and for sport or commercial fishing *(63)*.   More effective attractants may help to shorten the time required for young fish to become "rapid feeders." Several species of fish such as perch, the Chinese and Indian carps, and the European tench are very slow feeders, and an effective attractant should improve the rate of production.

DISEASE CONTROL   A wide range of chemicals has been used to control fish diseases, but presently only salt, acetic acid, and sulfamerazine are approved by FDA for all varieties of fish being raised for food *(62)*.   Use of oxytetracycline is restricted to trout, salmon, and catfish. Sulfa drugs and antibiotics have been used on some varieties for many years.   There is interest in the nitrofurans, first used in Japan to control bacterial fish diseases.   However, development of resistance to drugs may be a problem.

Chemicals as Research Tools

Chemical markers have been used to induce a permanent physiological mark in various species of fish and can be detected in larger fish by fluorescence examination.   Certain chemicals have been used to study physiological processes in various fish tissues.   Many radioactive isotopes have been incorporated into nutrients, intermediates, or environmental chemicals to monitor a particular biochemical pathway or to identify a particular site for accumulation of some physiologically active compounds.   These biochemical or physiological markers are valuable tools that could be used more in the sciences of fish physiology, nutrition, and biochemistry to identify or to confirm processes that help to determine growth, survival, or reproduction in fish populations *(67)*.

Outlook

There are a number of ways in which chemistry can contribute to the future growth of aquaculture *(74)*:

- A major bottleneck to large-scale and effective husbandry of many species is the lack of a satisfactory larval feed.   New technology is needed to make particles in fish feeds small enough for larval forms to ingest and use.

- Improved antioxidant systems are needed to extend life of feeds for fish. Attractants to make feeds more acceptable to fish could increase rate and efficiency of production.
- Disease control might be improved through use of new drugs.
- Sex-controlling chemicals might be useful for improving fish quality through delaying maturation in certain cases and in promoting more efficient use of water biospheres in which several species of fish live together.
- New technology is required for the tropic and subtropic regions of the world. Little research has been done on controlled breeding and rearing of most of the brackish water species found in those areas.

REFERENCES, CHAPTER 2

1. "1977 Handbook of Agricultural Charts," U.S. Dept. Agric., Agric. Handb. **1977**, No.524, and handbooks from preceding years. Preliminary estimates for 1978 by the Department.
2. "Agricultural Statistics," annual reports from U.S. Department of Agriculture, 1978, 1967, U.S. Government Printing Office: Washington, DC.
3. Estimates by Economics and Market Research Section, Tennessee Valley Authority, Muscle Shoals, AL.
4. Bibliographies on Fertilizer Manufacturing Processes and Use; prepared by Technical Library, Tennessee Valley Authority, National Fertilizer Development Center, Muscle Shoals, AL.
5. Harre, Edwin A. "What's Ahead in Fertilizer Supply–Demand"; presented at the *TVA Fertilizer Conference*, Cincinnati, OH, July 1976.
6. "The Waiting Game: Sizable Gas Reserves Untapped as Producers Await Profitable Prices"; *The Wall Street Journal*, February 3, 1977, pp. 11, 19.
7. "U.S. Oil Reserves Fell 1.7 Billion Barrels in '76, a Near Record, Trade Group Says"; *The Wall Street Journal*, April 8, 1977, p. 8.
8. Van Dyne, Donald L.; Gilbertson, Conrad B. "Estimating U.S. Livestock and Poultry Manure and Nutrient Production." U.S. Department of Agriculture, ESCS-12, p. 7.
9. "Marginal Deposits Will Satisfy Future Phosphate Demand," *Eur. Chem. News* **1977**, *30*(775), 22.
10. Davis, Charles H. "Energy Requirements for Alternative Methods for Processing Phosphate Fertilizers," *Proc. ISMA Tech. Conf.* Prague, Czechoslovakia Sept. 23–24, 1974, paper PTE/74/15.
11. Hurst, Fred J.; Arnold, Wesley D; Ryon, Allen D. *Chem. Eng. (N.Y.)* **1977**, *84*(1), 56–57.
12. MacLean, J.E. "NPK—Emphasis Potash"; *AIM 200–1976 NFSA Round-Up Proceedings Book* National Fertilizer Solutions Association: Peoria, IL, 1976; pp. 11–14.

13. U.S. Department of Agriculture, *Commercial Fertilizers, Statistical Reporting Service, Annual Reports.*

14. Davis, Charles H; Blouin, Glenn M. "Energy Consumption in the U.S. Chemical Fertilizer System from the Ground to the Ground," in *"Agriculture and Energy"*; Lockeretz, William, Ed.; Academic: New York, 1977; pp. 315–331.

15. Kilmer, V.J. "Nitrogen, Phosphorus, and Soil in Runoff from Agricultural Lands: An Overview," presented at 50th Annual Water Pollution Control Federation Conference, Philadelphia, PA, October 1977.

16. Mortvedt, J.J.; Giordano, P.M. "Crop Uptake of Heavy Metal Contaminants in Fertilizers," in *Biological Implications of Heavy Metals in the Environment"*; Energy Research and Development Administration Rep. Conf. 750929, Oak Ridge, TN, pp. 402–416.

17. Smith, A.E.; Secoy, D.M. "A Compendium of Inorganic Substances Used in European Pest Control Before 1850," *J. Agr. Food Chem.* **1976**, 24, 1180–1186.

18. Goring, C.A.I. "Prospects, Problems for the Pesticide Manufacturer," *Farm Chem.* **1976**, *139*(1), 18–26.

19. Goring, C.A.I. "The Costs of Commercializing Pesticides," in *"Pesticide Management and Insecticide Resistance"*; Watson, D.L., Brown, A.W.A., Ed.; Academic: New York, 1977.

20. "Pesticide Residues: Report of the 1967 Joint Meeting of the FAO Working Party and the WHO Expert Committee," *W.H.O., Tech. Rep. Ser. 391*, FAO Meeting Report No. PL/1967/M/11.

21. U.S. International Trade Commission, "Synthetic Organic Chemicals: United States Production and Sales, 1977," USITC Publ. 920, U.S. Government Printing Office: Washington, DC, 1978, and earlier editions.

22. "How Much Do Weeds Really Cost?" *Farm Chem.* **1978**, *141*(3), 122, 124.

23. Alder, E.F.; Wright, W.L.; Klingman, G.C. "Man Against Weeds," *Chem. Technol.* **1977**, 7(6), 374–380.

24. Hayes, W.J., Jr.; Vaughn, W.K. "Mortality from Pesticides in the United States in 1973 and 1974," *Toxicol. Appl. Pharmacol.* **1977**, 42, 235–252.

25. "Report of the First Session of the FAO Panel of Experts on Integrated Pest Control," FAO, Rome, 1967.

26. Smith, R.F. "History and Complexity of Integrated Pest Management," in "Pest Control Strategies"; Smith, E.H., Pimentel, D., Eds.; Academic: New York, 1978; pp. 41–53.

27. "Pest Control: An Assessment of Present and Alternative Technologies," in "Contemporary Pest Control Practices and Prospects: The Report of the Executive Committee"; National Academy of Sciences: Washington, DC, 1975; Vol. I, pp. 380–381.

28. Fowler, D.L.; Mahan, J.N. "Pesticide Review 1976"; Agricultural Stabilization and Conservation Service, U.S. Department of Agriculture: Washington, DC, 1977.

29. "Pesticides in the Modern World: A Symposium Prepared by Members of the Cooperative Programme of Agro-Allied Industries with FAO and Other United Nations Organizations," FAO, Rome, 1972.

30. "A Look at World Pesticide Markets," *Farm Chem.* **1979**, *142*(9), 61–68.

31. Hardy, R.W.F. In "Report of the Public Meeting on Genetic Engineering for Nitrogen Fixation"; U.S. Government Printing Office: Washington, DC, 1977; pp. 77–106.

32. "Agriculture Production Efficiency," National Research Council; National Academy of Sciences, Washington, DC, 1975.

33. Wittwer, S.H. "Food Production: Technology and the Resource Base," *Science* **1975**, *188*, 579–684.

34. Hardy, R.W.F.; Havelka; U.D.; Quebedeaux; B. "Carbon Assimilation"; Siegelman, H.W., Hind, G., Eds. Plenum: New York, 1978; pp. 165–178.

35. "Supporting Papers: World Food and Nutrition Study"; National Research Council; National Academy of Sciences: Washington, DC, 1977; Vol. 1.

36. "World Food and Nutrition Studies: The Potential Contribution of Research"; Report of the Steering Committee, NRC Study on World Food and Nutrition, and Commission on International Relations, National Research Council; National Academy of Sciences: Washington, DC, 1977.

37. Day, P.R. "Plant Genetics: Increasing Crop Yield," *Science* **1977**, *197*, 1334.

38. Zelitch, I. "Photosynthesis and Plant Productivity," *Chem. Eng. News* **1979**, *57*(6), 28–48.

39. Zelitch, I. "Improving the Efficiency of Photosynthesis," *Science* **1975**, *188*, 626–638.

40. "Crop Production—Research Imperatives"; Proceedings of an International Symposium, Boyne Highlands, MI, October 20–24, 1975, sponsored by Michigan Agricultural Experiment Station, East Lansing, MI; and Charles F. Kettering Foundation, Yellow Springs, OH.

41. Wittwer, S.H. "Alternatives Available for Improving Plant and Animal Resources," presented at the World Food Conference of 1976, Iowa State University, Ames, IA, June 1976, journal article No. 7693, Michigan Agricultural Experiment Station.

42. Bassham, J.A. "Increasing Crop Production Through More Controlled Photosynthesis," *Science* **1977**, *197*, 630–638.

43. Hardy, R.W.F.; Havelka, U.D. "Nitrogen Fixation Research," *Science* **1975**, *188*, 633.

44. "Effect of Increased Nitrogen Fixation on Stratospheric Ozone," CAST 1976, Report No. 53.

45. Evans, H.J. In "Report of the Public Meeting on Genetic Engineering for Nitrogen Fixation"; U.S. Government Printing Office: Washington, DC, 1977; pp. 61–67.

46. Lamborg, M.R. In "Report of the Public Meeting of Genetic Engineering for Nitrogen Fixation"; U.S. Government Printing Office: Washington, DC, 1977; pp. 56–70.

47. Carnahan, J.E., et al. "Nitrogen Fixation in Cell-Free Extracts of *Clostridium pasteurianium*," *Biochim. Biophys. Acta* **1960**, *38*, 188–189.

48. Hardy, R.W.F.; Havelka, U.D. In "Biological Solar Energy Conversion"; Mitsui, A., et al., Ed.; Academic: New York, 1977; pp. 299–322.

49. La Rue, T.A.; Kurz, W.G.W.; Child, J.J. In "World Soybean Research"; Hill, L.D., Ed.; Interstate Printers and Publishers, Inc.: Danville, IL, 1976; pp. 164–169.

50. Nickell, L.G. "Plant Growth Regulators," *Chem. Eng. News* **1978**, *56*(41), 18–34.

51. Hardy, R.W.F. In "Plant Regulation and World Agriculture"; Scott, Tom K., Ed.; Plenum: New York, 1979; pp. 165–206.

52. "World Food and Nutrition Study: The Potential Contribution of Research"; National Academy of Sciences: Washington, DC, 1977.

53. Van Demark, N.L., et al. "Increased Productivity from Animal Agriculture," Cornell University Report to National Science Foundation, Washington, DC, 1976.

54. Cunha, T.J. "Ruminant Production in Increasing Animal Foods in Latin America," in "Nutrition and Agricultural Development"; Scrimshaw, N.S., Behar, Moises, Eds.; Plenum: New York, 1976; pp. 355–360.

55. "Urea and Other Non-Protein Nitrogen Compounds in Animal Nutrition"; National Academy of Sciences: Washington, DC, 1976.

56. "Hormonally Active Substances in Foods: A Safety Evaluation," CAST Task Force Report, **1977**, *4*(1), 4–12.

57. Eggert, R.G. "Economic Benefit of the Use of Antibiotics in Animal Feed," report presented to the Subcommittee on Antibiotics in Animal Feeds of the National Advisory Food and Drug Committee, 1976.

58. Gillian, H.C.; Martin, J.R. "Economic Importance of Antibiotics in Feeds to Producers and Consumers of Pork, Beef and Veal," *J. Anim. Sci.* **1975**, *40*, 1241.

59. "Drugs in Livestock Feed," Technical Report, Office of Technology Assessment, Congress of the United States. U.S. Government Printing Office: Washington, DC, 1979; Vol. I.

60. "Agricultural Production Efficiency," National Academy of Sciences: Washington, DC, 1975.

61. Byerly, T.C. "Ruminant Livestock Research and Development," *Science* **1977**, *195*, 450–456.

62. "Aquaculture in the United States," National Research Council; National Academy of Sciences: Washington, DC, 1978.

63. "Proceedings First International Conference in Aquaculture Nutrition"; Price, K.S.; Shaw, W.N.; Danberg, K.S., Eds.; University of Delaware Press: Newark, 1976.

64. Smith, R.R.; Rumsey, G.L.; Scott, M.L. "Heat Increment Associated with Dietary Protein, Fat, Carbohydrate and Complete Diets in Salmonids: Comparative Energetic Efficiency," *J. Nutr.* **1978**, *108*(6).

65. David, H.S. "Culture and Diseases of Game Fish"; University of California Press: Berkeley, 1956.

66. "Manual of Fish Culture," U.S. Commission of Fish and Fisheries, Government Printing Office: Washington, DC, 1900.

67. "Fish Nutrition"; Halver, J.E., Ed.; Academic: New York, 1972.

68. Halver, J.E. "Formulating Practical Diets for Fish," *J. Fish Res. Board Can.* **1976**, *33*, 1032–1039.

69. Halver, J.E.; Smith, R.R.; Tolbert, B.M.; Baker, E.M. "Utilization of Ascorbic and Acid in Fish," *Ann. N.Y. Acad. Sci.* **1975**, *258*, 81–102.

70. "Nutrient Requirements of Trout, Salmon and Catfish," National Research Council; National Academy of Sciences: Washington, DC, 1973.

71. "Nutrient Requirements of Warmwater Fishes," National Research Council; National Academy of Sciences: Washington, DC, 1977.

72. Sinnhuber, R.O. "The Role of Fats," in "Fish in Research"; Neuhaus, O.W., Halver, J.E., Eds.; Academic: New York, 1969.

73. "Report of the 1970 Workshop of Fish Feed Technology and Nutrition," U.S. Fish and Wildlife Service, Resource Publ. 102, 1970.

74. "Fin Fish Nutrition and Fish Feed Technology"; Halver, J.E.; Tiews, K.; Eds.; Heeneman: Berlin, 1979.

Handling, Storing, Preserving, and Processing of Foods

Revolutionary changes have taken place in the U.S. food system in this century. Entirely new foods line the grocery shelves—convenience foods designed to reduce the time spent in preparation in the home, and foods specifically for those with special dietary needs. The traditional foods are still there, but in some cases in more nutritious forms—milk fortified with vitamin D, and bread in which some of the vitamins and minerals removed in milling have been restored. Fresh fruits and vegetables once on the market for brief periods of time are now there year-round. Without additives, this great variety of foods would not be available.

In the United States today, very little food is eaten as it comes from the field. Even the farm family relies on the grocery store for much of what it eats. The nation is served by a large, industrialized food system that has been able to supply most Americans with abundant amounts and a wide assortment of foods at reasonable prices. For some years, Americans have spent 16–17% of their income for food (Table 3.1), which is somewhat below that spent in a number of industrialized nations (Figure 3.1).

ADDITIVES IN FOOD

The additives used by the food industry serve a number of purposes. They:

- Facilitate processing, handling, distribution, and preparation in the home.

TABLE 3. 1   Food Expenditures as Percentage of Income, 1960-78

Year	Food Expenditures as Percentage of Income
1960	20.0
1961	19.8
1962	19.3
1963	18.9
1964	18.4
1965	18.1
1966	18.0
1967	17.2
1968	16.9
1969	16.4
1970	16.2
1971	15.7
1972	15.7
1973	15.9
1974	16.8
1975	16.9
1976	16.8
1977	16.7
1978	16.5[a]

Source: Ref. *53*.
[a] Preliminary.

- Control chemical, physical, and microbiological changes so as to reduce waste, lessen hazards from microorganisms, and preserve quality.   (Such use of additives is an ancient practice.   Early in history, people learned to preserve foods by smoking them or by adding salt, sugar, vinegar, or spices).
- Extend shelf life.
- Improve sensory and nutritive properties.

Kinds of Additives

The Food Protection Committee of the Food and Nutrition Board (National Research Council–National Academy of Sciences) defines a food additive as "a substance or mixture of subtances other than a basic foodstuff, which is present in a food as a result of any aspect of production, processing, storage, or packaging" *(1)*.   There are two broad types: direct additives, which are added deliberately to perform specific functions such as to improve flavor or nutritive value; and indirect additives, which are substances present in foods in trace quantities as a result of some phase of production, processing, storage, or packaging.

Foods may also contain various natural components (for example, safrole and related compounds in sassafras root, and antithyroid compounds in certain plants) and contaminants (for example, aflatoxin, a fungal toxin) known to be toxic to man *(2)*.   Because some natural components, as well as indirect additives, can be highly potent, serious effort should be made to identify, monitor, and, when necessary, regulate them.

Many of the additives used by the food industry come from natural sources. For example, the lecithin in bread sold in supermarkets (added to improve volume, uniformity, and fineness of grain) is extracted from soybeans and corn.   In some cases where natural sources cannot provide adequate supplies, the substances can be synthesized to supplement natural supplies.   For instance, vanillin, the main flavoring substance in vanilla beans, is synthesized at a cost considerably lower than the cost of the natural product.   It is doubtful that natural sources of vanillin could meet consumer demands. The vitamins and minerals used to improve the nutritive value of certain

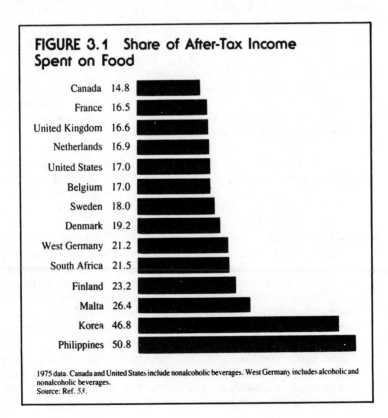

FIGURE 3.1   Share of After-Tax Income Spent on Food

Canada	14.8
France	16.5
United Kingdom	16.6
Netherlands	16.9
United States	17.0
Belgium	17.0
Sweden	18.0
Denmark	19.2
West Germany	21.2
South Africa	21.5
Finland	23.2
Malta	26.4
Korea	46.8
Philippines	50.8

1975 data. Canada and United States include nonalcoholic beverages. West Germany includes alcoholic and nonalcoholic beverages.
Source: Ref. 53.

foods are manmade; they are identical in composition to the vitamins and minerals found naturally in food but are generally less expensive.   Compared to their natural counterparts, manmade additives often are purer, more concentrated, and more consistent in quality.

Overall, foods or food components from natural sources constitute well over 99% of the weight of the American diet *(2)*.   Food additives comprise the bulk of the balance and are usually either dietary supplements such as vitamins and minerals, or materials derived, or derivable, from natural sources.   Pesticide residues and contaminants of both natural and manmade origin contribute only trace amounts to the American diet.   Thus, only a small fraction of 1% of the American diet is not derived from ordinary foods *(2)*.

Factors Affecting Use of Additives

The use of additives in the U.S. food industry has grown hand in hand with a number of national developments:

- *Population growth.*   As the population has grown, more land has been devoted to nonagricultural purposes such as housing and roads, greatly increasing the cost of land.   Thus it is important to maximize food production from the land that is still available for farming, and to minimize the waste that can occur during harvest, processing, storage, and distribution.   In addition, new foods produced from unconventional sources, such as petroleum and wastes, would help offset the scarcity of agricultural land.
- *Urbanization.*   The steady movement of Americans from farms to cities has increased the mean distance between the point of food production and the point of consumption.   This, in turn, has increased distribution time and has resulted in a retail marketing system involving very large inventories.   Thus, food must be preserved, which usually requires chemical additives.
- *Increasing labor costs.*   The higher wages now being paid to agricultural workers encourage farmers harvesting plant foods to use chemical additives that control the rate and uniformity of ripening and ageing.   Uniform ripening and ageing assist substantially in the commercial success of "once-over" mechanical harvesting, which increases productivity per man-hour of labor.   In their efforts to hold the line on food costs, other segments of the food industry are also motivated to use chemical additives that reduce labor costs.
- *Public health concerns.*   Adding chemical nutrients to certain foods has, in some instances, improved the health of the American people.   For example, the addition of vitamin D to milk and potassium iodide to salt has greatly reduced the incidences of rickets and goiter.

- *Consumer desires.* The increase in consumer demand for processed convenience foods is readily apparent from their rapidly increasing sales. This demand has intensified as more women have entered the job market and the earning power of families has increased. Additives are essential components of most convenience foods. In addition, there is considerable demand by consumers for a large variety of high-quality fresh foods on a year-round basis. Meeting this demand efficiently requires a broad range of additives. For example, many fresh fruits and vegetables are treated with sulfur dioxide gas before shipping. This procedure extends storage life so that the product can be either exported during the season or imported during the off season. The useful life of potatoes can be extended by applying a chemical that inhibits sprouting, and some vegetables, such as cucumbers and rutabagas, and fruits, such as apples, can be coated with wax to retard moisture loss and produce a bright appearance. Also, waste-free items can be marketed. For example, apples can be peeled, cored, and sliced, and the slices preserved by dipping them in a bisulfite, or combination bisulfite dip, before packaging.
- *Special dietary needs.* Increasing attention is being given to population groups with special dietary needs. Thus, foods low in salt, fat, sugar, or cholesterol are now available in many food stores. Preparation of these special foods, with characteristics suitable for the U.S. marketing system, would, in most instances, be impossible without the use of additives. For example, a large amount of sugar is required in regular jams and jellies. However, to produce a low-calorie product, a low-methoxyl pectin (or pectin plus a gum) must be added to give the desired consistency. Also, nonnutritive sweeteners are required in the manufacture of low-calorie or sugar-free products such as beverages and puddings.

Regulation of Food Additives

The use of additives in food is regulated by the Food, Drug, and Cosmetic Act and its 1954–60 amendments. (For further discussion of the Act, see Chapter 5). To receive approval to use an additive, the manufacturer or industrial user must first prove that it is safe. FDA approves the use of an additive if it meets these basic criteria:

- The quantity to be used does not exceed the amount reasonably needed to accomplish the intended effect.
- The amount to be used has an adequate margin of safety based on animal studies and other safety information.
- The substance is of appropriate purity for use in food.
- Use of the substance does not promote deception.
- The substance will not induce cancer when ingested by man or animal.

Some people believe that excessive amounts of additives are used in our foods. Their concern is that small quantities of additives (and other environmental contaminants as well) form part of a hazardous background that contributes to chronic diseases. It is true that by choosing carefully, a person can maintain a healthful state on a diet essentially free of food additives, as some of our ancestors and a small fraction of our current population have demonstrated. The question, then, is whether it is feasible or desirable in the United States, either now or in the future, to largely free our food supply of intentional chemical additives, as defined in the broadest (nonlegal) sense.

The answer is yes—if the public wants this to occur and is willing to accept the disadvantages that would accompany such a move. In many cases, additives provide a product with improved qualities at a lower overall cost. For example, bread made without additives can be kept only for a day or so before it becomes stale, although it is still nutritious. In the United States, most bread contains various additives that delay spoilage and help to maintain a soft texture for several days. According to a study made in 1973, bread without additives cost 17% more than bread with additives *(3)*. The increase resulted primarily from shorter shelf-life, which increases distribution and selling costs. The increase is probably even greater today because of higher energy costs *(4)*.

The study estimates that in 1973, removing additives from bread, margarine, processed meats, and processed cheese would have cost a family of four about $50 per year for purchase of those four foods, which contribute only about one-fourth of the total caloric intake. Thus if use of additives is to be greatly curtailed, the American consumer must be willing to pay more for food, be content with less convenient forms of foods, do without some foods entirely, and lose various other advantages resulting from the use of chemical additives.

HANDLING (TRANSPORTING AND PREPROCESSING) OF FOODS

Foods, whether plant or animal, grow and reach a point of optimum quality for human consumption. Then, after harvest or slaughter, certain chemical processes continue. Eventually, fresh foods will be decomposed by microorganisms, unless their growth is prevented.

Chemicals are applied to regulate the growth of fruits and vegetables during and immediately after harvest (Table 3.2). For example, for about 20 years a plant growth regulator, gibberellic acid, has been widely used to retard ripening of tomatoes, bananas, and guava, thus facilitating their distribution and extending their marketable life.   Applying this chemical to navel oranges and lemons keeps them on the tree longer *(5)*.   In the case of oranges, the peel is made firm, the degradation of chlorophyll is delayed, and the accumulation of carotenoids, one of which is a precursor of vitamin A, is delayed.   In the case of lemons, the level of soluble solids such as sugar and vitamin C is increased.

While some plant growth regulators delay ripening, others hasten it to meet market demand and also to obtain uniform ripening.   For the past 25 years or so, minute quantities of ethylene gas have been applied to tomatoes,

TABLE 3.2   Chemicals Applied by the Food Industry to Control Properties of Fruits and Vegetables During and Immediately Following Harvest

Specific Objective	Chemicals Used	Commodities
Delay ripening	Gibberellic acid	Tomato, guava, banana, koki fruit, lemon, orange
	Maleic hydrazide	Mango, tomato
	Cycloheximide	Pear
	Vitamin K_1 and K_3	Banana
	Maleic acid	Banana, citrus
	Ethylene oxide	Mango, tomato
	Sodium dehydroacetate	Strawberry
	Potassium permanganate	Banana
Hasten ripening	Ethephon and ethylene	Tomato, pineapple, cantaloupe, fig, peppers, banana, citrus
	Abscisic acid	Orange, banana
	Acetylene	Banana, tomato, lemon, orange
	Oil	Fig
	2, 4-D	Banana, sapota, guava
Reduce weight loss	Chlorophenoxy acetic acid	Mandarin orange, bean
	β-Naphthoxyacetic acid	Mandarin orange
Decrease loss of vitamin C	Gibberellic acid	Citrus
	Chlorophenoxy acetic acid, β-Naphthoxyacetic acid	Mandarin orange
Retain green color	Chlorophenoxy acetic acid	Bean
	2, 4-D	Green vegetables
	Cyocel	Lettuce
	Cycloheximide	Pear
Control firmness	Gibberellic acid	Citrus
	Alar	Apple
	Ethylene oxide	Mango

bananas, and other fruits, increasing their respiratory rate and hence stimulating ripening.   Ethephon (2-chloroethylphosphonic acid), an ethylene-releasing compound, is easier to handle, and is now used on a large portion of tomatoes grown in the United States.   Ethephon treatment of tomatoes has the following advantages:

- Sorting cost is reduced because of uniform ripening.
- Weight loss is reduced and shelf life is prolonged.
- Ripening rooms are not necessary.
- Yield is increased for once-over harvest.
- Maturity is hastened so that fruit can be marketed earlier in the season.

Other chemicals have been shown to delay deterioration of tangerines in storage.   When tangerine leaves are treated with chlorophenoxyacetic acid and β-napthoxyacetic acid before harvest, weight loss from the fruit is reduced during storage, thereby maintaining nutrient quality.

TABLE 3.3   Chemicals Applied by the Food Industry to Increase Storage Life of Fresh Fruits and Vegetables

Specific Objective	Chemicals Used	Commodities
Inhibit micro-organisms	Sulfur dioxide	Stone fruit, strawberry, citrus, grape
	Sodium bisulfite	Grape
	Sulfur	Peach
	Thiourea	Orange
	Zinc salt of pyridine-N-oxide-2-thione	Peach
	Thiobendazole	Pear
	Biphenyl	Citrus, grape, peach, potato
	8-Hydroxyquinoline	Strawberry
	Sodium salicylanilide	Banana, orange
	Ammonia	Citrus, peach
	2-Aminobutane	Citrus, apple, peach
	Captan	Strawberry, cherry, pear
	2, 7-Dichloro-4-nitroaniline	Sweet cherry, peach
	Benomyl	Tomato
Reduce weight loss	N^6-Benzyladenine	Green vegetables
	Naphthalene acetic acid	Cauliflower, cabbage
Inhibit sprouting	Maleic hydrazide	Onion, radish, turnip, carrot, potato
	Isopropyl-N-(3-chlorophenyl) carbamate	Potato
Reduce leaf abscission	Napthalene acetic acid	Cauliflower, cabbage

STORING FRESH FOODS

Storing fruits and vegetables in the fresh state has enabled Americans to enjoy delicacies year-round that a generation ago were available for only a few weeks.   But during storage, a variety of deleterious changes can occur. The major portion of the losses in fresh fruits and vegetables in storage results from the action of fungi and bacteria; however, many other undesirable physical and physiological changes can also occur: green vegetables lose their color and shrivel, onions and potatoes sprout, and cabbage heads lose leaves (Table 3.3).   About 40% of the produce harvested in the United States is never consumed because of spoilage.   Losses amount to over $10 billion annually.   Treatments with plant growth regulators and other chemicals are helping to reduce these losses.

The most serious infestations during storage are those that cause rapid and extensive breakdown of high-moisture commodities, sometimes spoiling an entire bushel or crate.   This condition is typified by attacks of a fungus (*Rhizopus* sp.) on stone fruits and strawberries, a mold (*Penicillium* sp.) on citrus and pome fruits, and a bacterium (*Erwinia carotovora*) on leafy vegetables and potatoes.   All can be controlled by appropriate chemical additives.   One of the additives most widely used on fresh fruits is sulfur dioxide, which is effective even at a low concentration in killing microorganisms.   This treatment, in combination with reduced temperature, maintains the attractive appearance and good flavor and texture of fresh fruits for a considerable time, making them available in remote areas that could not ordinarily be served.

It is standard commercial practice to fumigate stored grapes every ten days with sulfur dioxide to prevent the spread of a fungus (*Botrytis*) (6). (This use of sulfur dioxide has been known since the time of the ancient Egyptians and Romans, who apparently used fumes from burning sulfur as a sanitizing agent in making wine.)   To prevent growth and spread of the disease in packaged grapes during storage and long distance shipping, a sodium bisulfite packet, which gives off sulfur dioxide gas slowly, is sometimes added.

New growth regulators that may yield still further improvements in storage of fresh produce are being investigated.   Among the more promising are N^6-benzyladenine and 6-aminopurine.   Added at concentrations of only a few parts per million, compared to the presently used captan which must be added at a 1% level, they enable a number of vegetables—including lettuce, beans, onion, broccoli, celery, and asparagus—to remain green longer (7,8).   The growth regulators can be sprayed on the vegetables in the field, which is more economical than dipping, or other methods used in processing plants. Two plant hormones, zeatin and dihydrozeatin, also appear promising. A single 100-ppm treatment more than doubled the refrigerated storage life of broccoli *(9)*. These compounds seem to aid chlorophyll retention, alter respiration rate, retard protein degradation, and increase the formation of organic acids. FDA has not yet set tolerances for the use of most of these compounds.

PRESERVING AND PROCESSING OF TRADITIONAL FOODS, INCLUDING CONVENIENCE FOODS

Originally, simple preservation procedures such as heating or drying were used to stabilize food for later consumption. Both procedures, however, often result in undesirable changes in flavor, color, texture, and nutritive value *(10)*. It has become evident that modern food technology can provide a greater variety of more nutritious and flavorful products by combining food-grade chemicals with traditional preservation procedures (Table 3.4). The type of chemical used and its specific function vary with the particular preservation method involved.

Chemicals Used During Thermal Processing

Using heat to pasteurize or sterilize a food product is one of the most important preservation methods available. However, excessive heating will degrade the product. Carefully selected chemical additives permit short-time, high-temperature processing that can minimize the undesirable effects of heating.

Acids and salts Incorporating acids or salts during thermal processing results in a more palatable and flavorful product that will maintain these qualities during normal distribution and storage.

Tomatoes are one of the largest volume products canned commercially in the United States, with over 40 million cases packed annually. Some canning tomatoes, particularly in southern California, are occasionally not acidic enough, possibly because they were overripe or watered excessively in very warm growing areas. This low acidity necessitates increased processing times to achieve sterilization. To avoid increased processing times, some canners add citric acid to increase the acidity. Other vegetables, such as artichoke hearts, are also acidified with acids such as citric or acetic for this same purpose *(11)*.

TABLE 3.4   Some Chemicals Permitted for Use in Traditional Processed Foods

General Purpose	Chemical	Specific Function	Product
Preserve quality	Propionates	Retard molds	Cheese, baked goods
	Sodium benzoate	Retard bacteria, yeasts	Soft drinks, acid foods
	Sorbic acid, potassium sorbate	Retard molds, yeasts	Acid foods, cheese, wines
	Sulfur dioxide	Retard all microorganisms, inhibit browning	Dried fruit, wines, fruit slices, grapes
	Acetic acid (including vinegar), acetates	Retard bacteria, yeasts	Acid foods, baked goods
	Sucrose	Inhibit microorganisms	Jam, jelly
	Carbonyls, organic acids	Retard microorganisms	Fish, meats
	Esters of p-hydroxybenzoic acid	Retard molds, yeasts	Beverages, artificially sweetened jelly
	Nitrite (future use under study)	Inhibit botulinum growth	Cured meats
	Sodium chloride (salt)	Control microorganisms, retard enzymatic browning	Fish, meat, pickles, fruit slices
Stabilize or improve nutritional content	Butylated hydroxyanisole (BHA), butylated hydroxytoluene (BHT), α-tocopherol	Stabilize vitamin A, inhibit oxidation	Fish products, sausages, potato chips, nut meats
	EDTA, citric acid, sulfur dioxide	Stabilize vitamin C	Black current juice, beverages
	Sodium chloride (salt)	Retain water-soluble nutrients	Frozen fish
Stabilize or improve texture	Calcium compounds	Increase firmness	Canned and frozen fruits and vegetables
	Rennin	Coagulate casein	Cheese
	Papain	Tenderize	Meat
	Sodium hexameta-phosphate	Soften texture	Canned peas, beans, meat, poultry

TABLE 3.4   Some Chemicals Permitted for use in Traditional Procecessed Foods (continued)

General Purpose	Chemical	Specific Function	Product
Stabilize or improve texture	Sodium acid pyrophosphate	Leaven	Bakery products
	Polyoxethylene sorbitan esters	Emulsify	Ice cream, cake mixes, shortening
	Glycerol	Promote water retention	General use
	Calcium silicate	Prevent caking	Salt, baking powder
Stabilize or improve flavor	Ascorbic acid	Prevent oxidation	Canned foods
	BHA, BHT, propylgallate, α-tocopherol	Prevent oxidation and browning	Potato granules, nut meats, diced meats
	Citric acid	Prevent oxidation	Canned fruits and vegetables
	Triethyl citrate	Chelate metals and lessen oxidation	Egg white
	Citrates of calcium or potassium	Chelate metals and lessen oxidation	Oils, fats, salad dressing
	Organic acids	Enhance flavor	Carbonated beverages
	Sodium chloride	Enhance flavor, retard browning	General use, cut fruits
	Synthetic flavors	Enhance flavor	General use
	Sweeteners	Enhance flavor	General use
	Lecithin	Enhance flavor	Oleomargarine, shortening, candies
	Monosodium glutamate	Enhance flavor	General use
	Maltol	Enhance flavor	Soft drinks

TABLE 3.4   Some Chemicals Permitted for use in Traditional Processed Foods
(continued)

General Purpose	Chemical	Specific Function	Product
Stabilize or improve color or appearance	Sulfur dioxide	Prevent browning	Dried fruits, potatoes, wines
	Ascorbic acid	Prevent enzymatic browning	Frozen fruit slices
	Sodium chloride	Prevent enzymatic browning	Apple slices
	Sucrose	Prevent enzymatic browning	Fruit slices
	Citric acid & malic acid	Retard browning	Fruit slices
	EDTA (disodium)	Prevent discoloration by inactivation of trace metals	Potatoes, corn
	Beet powder	Provide red color	Various foods
	Annatto	Provide yellow color	Oleomargarines, shortening
	Nitrites (future use under study)	Alter and fix color	Meats
	Polyvinylpyrrolidone	Clarify	Beverages, vinegar
	Pectinase	Clarify	Fruit juices

Acids are added for other reasons.   Malic acid is sometimes used as a color stabilizer in apple, grape, and other fruit juice drinks.   Furthermore, the stability of some vitamins is dependent on pH.   Citric acid serves to inhibit development of off-flavors in foods that contain fats or oils, as well as to retard loss of color (antibrowning) and flavor in some canned fruits and vegetables (12).

The textural properties of canned tomatoes and potatoes are improved by the addition of calcium salts before processing.   Also, calcium and sodium phosphates are used as buffers or stabilizers in many products such as canned potatoes, green beans, peppers, tomatoes, sauces, and evaporated milk, providing a firmer, crisper texture or a more brightly colored product (12).

Canned pears are more susceptible to developing a pink discoloration if traces of copper, iron, or zinc are present.   This discoloration can be eliminated by adding citrates or phosphates to bind the metal ions.   Similarly, these chemicals are added to some canned legumes to avoid darkening (12).   Disodium ethylenediaminetetraacetic acid (EDTA) is sometimes added to canned vegetables such as black eyed peas, kidney beans, potatoes, and corn, combining with the trace amounts of copper, iron, and chromium present, and so preventing the product from turning gray.   EDTA is also added to canned shrimp to retard color changes.

STABILIZERS AND EMULSIFIERS   The use of stabilizers and emulsifiers permits production of foods that would otherwise not be available to the consumer—for example, instant-type products that would require considerable time to prepare in the home.   Gums such as alginates are sometimes added to canned specialty items that contain butter or vegetable oils— gravies, for example—to increase the viscosity and retard separation of the oil.   Starches are added to canned foods such as baby foods, soup, sauces, and gravies to increase their viscosity so that the solids remain in suspension.   In canned pie fillings, gums are used as a partial or complete replacement for starch, which often fails to maintain a uniform liquid phase. Adding carrageenan to canned gelled dessert products renders them stable when they are not refrigerated.   The same gum is sometimes added to tomato juice to maintain an even suspension of the fine pulp and to improve its feel in the mouth.   Other gums are used to suspend natural fruit pulps in fruit juices and soft drinks.

NITRITES   Some foods are smoked, and a mild preservative action results from various chemicals in smoke, such as carbonyls and organic acids (13).   Thermal processing, which usually accompanies the smoking, inactivates microorganisms by the combined effects of drying and elevated temperatures. Sodium nitrate and sodium nitrite have, in the past, been added to smoked-cured meats to stabilize color *(1)* and, even more importantly, to protect against the growth of the toxic botulism organism. However, during heating and storage, some of the nitrite may react with other food constituents to produce extremely small amounts of nitrosamine, which is carcinogenic in rats *(14–17)*.

In 1973, USDA appointed an expert panel to study this problem, and the panel recommended that the use of nitrates, which can be converted to nitrites, be completely eliminated and that the use of nitrites be reduced to the lowest practical level. USDA and FDA have banned the use of nitrates and reduced the use of nitrites to levels at which the presence of nitrosamines cannot be detected. One way of reducing nitrites while still protecting against botulism is to use it with another antimicrobial chemical, potassium sorbate. In a more recent study for FDA at the Massachusetts Institute of Technology, nitrites themselves, rather than nitrosamine, were found to cause cancer in rats *(18)*. The results are being evaluated by the Universities Associated for Research and Education in Pathology under a one-year contract with FDA.

A concerted effort is underway on the nitrites problem in private and government laboratories. FDA is planning a three-year research program aimed at isolating and identifying the specific mechanism by which sodium nitrite inhibits botulism, with the aim of using this information to develop alternative materials. USDA is allocating about $1 million in its 1980 budget to develop an acceptable alternative. Meanwhile, sodium nitrite can still be used in reduced amounts.

Chemicals Used During Freezing

Freezing food preserves it by lowering the product temperature to about -10 to -20° C to slow down deleterious chemical reaction and to stop growth of microorganisms. Chemical additives are used in most frozen foods primarily to ensure that the product will not undergo significant chemical degradation during frozen storage. For instance, freezing does not inactivate enzyme systems. Unless they are inhibited in some manner, they are able to react slowly during frozen storage and rapidly following thawing. With vegetables, enzymes are inhibited by a mild heat treatment (blanching) that has no objectionable effects, since vegetables are normally cooked prior to consumption. However, fruits are not usually blanched because the process adversely affects flavor and texture. In most frozen fruits, enzymes are controlled by the addition of sulfites or ascorbic acid or by packing the fruit in sugar. Addition of sugar to fruits, aside from contributing to flavor, displaces oxygen from the container, thus reducing oxidative discoloration and oxidative loss of vitamin C and other nutrients *(19)*. Citric acid is also helpful in retarding oxidative reactions by inactivating trace metals that would otherwise promote oxidation. Calcium salts are occasionally added to frozen fruits and vegetables to give a firmer texture.

Meats, in general, are frozen without additives. However, fish are sometimes dipped in a trisodium polyphosphate solution prior to freezing to reduce loss of natural liquids during thawing *(19)*.

In some frozen dairy products, lactose can crystallize, resulting in an undesirable feel in the mouth and contributing to protein instability. Lactose crystallization can be inhibited by adding the enzyme lactase or by adding sucrose at a level of 5–10% *(20)*.

Among new convenience foods, frozen potato products are major items. Preparation of french fries involves treatment with either citric acid or a buffer such as sodium acid pyrophosphate to neutralize any alkali (added to remove the peel) that might remain following the peeling operation, and to ensure that the product will retain a desirable light color during frozen storage. Modified starch is needed to bind potatoes in the form of frozen balls. Sauces for frozen vegetables and frozen chicken, meat, and fish pies also require modified starch, since regular starch does not perform satisfactorily. In preparation of frozen dessert items, additives such as alginates, carrageenan, xanthan gum, gum acacia, or gelatin are often used to produce and maintain a satisfactory product texture during storage *(12)*.

Chemicals Used During Drying

Drying, probably the oldest method of food preservation, relies on the principle that microorganism cannot grow below a certain water content. However, even at low water contents, degradative chemical reactions can still occur, and these reactions must be minimized to maintain product quality. Sulfur dioxide or bisulfites are the most important chemicals used for this purpose. Prior to drying, cut fruits such as apricots, peaches, pears, and apples and vegetables such as cabbage, potatoes, and carrots are treated with sulfur dioxide, either as a gas or a dip. This treatment helps to maintain the original color of the product during processing and normal storage, to prevent spoilage from insects and microorganisms, and to protect vitamin C *(21)*.

In the United States, over 150,000 tons of prunes and figs are produced annually, and a large percentage of this crop is marketed in a semidry condition (over 28% moisture). This means the product must be pasteurized or treated with an antimicrobial compound such as potassium sorbate to prevent spoilage from mold and yeasts *(22)*.

Dried egg whites often contain additives such as sodium lauryl sulfate to enhance whipping properties, triethyl citrate to inhibit oxidation, and silicates and stearates to avoid caking. Dried fish products sometimes contain sulfurous salts to improve flavor *(23)*. Freeze-drying, while relatively expensive, offers special advantages in terms of preserving the fresh quality of foods. No additives are usually used in this process, except bisulfite in some fruits and occasionally an antioxidant (ascorbic acid).

Chemicals Used During Fermentation

Fermentation, used as a food processing tool, differs from microbial spoilage in that it involves growth of beneficial microorganisms and changes in the properties of the food, such that harmful microorganisms cannot grow. Sodium chloride is an important chemical in food fermentations, since its concentration in the product influences the types of microorganisms that will predominate. Vegetables such as cauliflower, onions, and green tomatoes are brined or salted using a 2–15% salt solution, the level favoring fermentation by lactic-acid-forming bacteria. Olives and cucumber pickles are prepared in a similar way, with acetic acid sometimes being added to help prevent the growth of undesirable microorganisms. Salt is added to cabbage to encourage the lactic acid fermentation required to produce sauerkraut, and sorbate to keep any surface yeast from growing.

Alcoholic beverages are produced by the growth of either natural microflora or, more commonly, by the growth of a prepared culture of microorganisms. In some wines, sulfur dioxide is added as a means of controlling growth of unwanted microorganisms.

Adverse Effects of Chemical Additives

Although food additives serve many valuable purposes, some of them also have specific disadvantages, aside from the bigger question of whether the amounts used in the food system may have long-term deleterious effects on health. (See section on "Assuring the Wise Use of Chemicals.") For example, citric acid added to tomato products increases tartness. This taste difference, however, can be minimized by adding a small amount of sugar *(24)*. In some instances, sorbates and propionates can result in undesirable flavor changes. Ascorbic acid accelerates nonenzymatic browning reactions and so should only be used in products where nonenzymatic browning is not a problem, such as in preserving color of frozen fruits.

Sulfur dioxide bleaches some pigments.   Furthermore, it causes rapid degradation of thiamin (vitamin B_1), results in softening of some plant tissue products, has an objectionable taste to sensitive individuals, and may cause metal cans to blacken.

There is a clear need to more fully understand the interactions that occur when chemicals are added to food.   Not fully understood, for example, is why sodium chloride, even at very low levels, reduces the sour taste in foods and beverages, and sugars reduce the intensity of both salty and sour tastes.   The use of additives such as salt, sugars, or fats for technical reasons in food processing also has public health implications, for they increase the dietary intake of these materials, which may be harmful to some segments of the population.   Even more important is the need to learn how to deal with any undesirable reactions that may occur.   In all instances, the greatest care must be taken to assure that chemicals used are safe under the conditions employed, from the point of addition to the point of consumption.

FORTIFICATION OF FOODS WITH VITAMINS, MINERALS, AND AMINO ACIDS

Until recently, consumers considered their nutritional needs in terms of foods like meat, potatoes, bread, milk, and eggs.   USDA nutritional guides (25, 26) for the consumer were built around daily selections of seven groups of foods.   For greater ease of understanding and use, the seven groups were later reduced to four—milk and dairy products; meat, poultry, and fish; vegetables and fruit; bread and cereal.

With the accumulation of knowledge over the past century, the specific nutrients required in food are, in the main, well-known.   Proper balance of both macronutrients [proteins (amino acids), carbohydrates, and fats] and micronutrients (vitamins and minerals) is required, and informed consumers have come to recognize that certain foods are associated with certain nutrients.

Some 45 nutrients are now known to be needed for proper nutrition, and "nutrient allowances" for man have been established for approximately half of them (27).   The nutrients have been identified by chemical structure, and their levels in various foods measured.   Knowledge of the chemical structure of the nutrients enabled the organic chemist to synthesize some of them, leading to industrial production.   As production has increased, costs have declined, as has been illustrated with vitamin A (Figure 3.2).

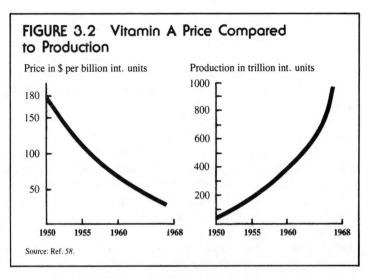

FIGURE 3.2   Vitamin A Price Compared to Production

Price in $ per billion int. units

Production in trillion int. units

Source: Ref. 58.

While an awareness of the importance of nutrition was evolving, marked changes were taking place in the food supply system.   Advances in food technology have broadened the selection of foods from which to choose, thus multiplying and complicating the choices for the consumer.   Now there is a selection of "convenience," "semblance," "fabricated," and "novel foods," some simulating known foods and others having no past counterpart, as well as the traditional, processed, and refined foods.   The number of foods and food ingredients in American supermarkets has risen to nearly 10,000.   This increasing variety, combined with changes in the nature of the family structure, the fast pace of changing food habits, and lack of adequate nutrition knowledge, complicates the problem of achieving proper nutrition.   Over the past decade, surveys of the U.S. population have indicated that some people are not receiving the recommended daily allowances of nutrients, particularly certain vitamins and minerals.   Clinical and biochemical studies have verified that some segments of the population do suffer from certain deficiencies—for example, iron deficiency, leading to anemia.   In developing countries, deficiencies are widespread— for instance, 250,000 children go blind each year from vitamin A deficiency.

The Evolution of Fortification

Adding nutrients to foods can maintain or improve the quality of a diet of a group, community, or population *(28, 29)*.   Various terms have been used to describe the process, including standardization, enrichment, and nutrification.   The Joint Committee on Nutrition of the Food and Agriculture

Organization, and the World Health Organization consider the term "fortification" to be the most appropriate *(30)*. The term was intended to apply principally to the use of relatively small quantities of nutrients, although it is difficult to draw a sharp line between that and some formulations of food mixtures containing, for example, moderate amounts of protein concentrates.

Adding to a food for nutritional improvement is not a new concept. Iodine (in the form of iodide) was added to salt in the 19th century in South America *(31)*. The 20th century saw fortification of vegetable-fat spreads with vitamin A in Europe *(32)*, and "restoration" of micronutrients to milled cereal products in Great Britain and the United States. In 1939, the Council on Foods and Nutrition of the American Medical Association (AMA) published a policy on the addition of specific nutrients to foods, followed in 1941 by one from the Food and Nutrition Board of the National Research Council (NRC). Fortification in 1940 of white flour and bread with three vitamins and iron, World War Food Order No. 1, was endorsed by both AMA and NRC. Drawing on all available information from various studies, the Food and Nutrition Board has been providing guidelines since the cereal "enrichment" program was developed in the 1940s. The current revision

TABLE 3.5   Vitamins and Deficiency Symptoms

Vitamin	Deficiency
Fat-Soluble	
Retinol (A)	Eye lesions (xerophthalmia, keratomalacia, or blindness)
Calciferol (D)	Bone malformations (rickets, osteomalacia)
α-Tocopherol (E)	Blood disorders (microcytic anemia and edema in prematures; creatinuria, red cell hemolysis)
Phylloquinone (K)	Hemorrhage, decreased clotting
Water-Soluble	
Thiamin (B_1)	Beriberi
Riboflavin (B_2)	Mouth, skin, and eye lesions
Niacin (B_3)	Pellagra
Folacin	Macrocytic anemia
Biotin (H)	Seborrheic dermatitis
Pantothenic acid	Gastrointestinal and nervous disorders
Pyridoxine (B_6)	Convulsions, dermatitis
Cobalamin (B_{12})	Pernicious anemia
Ascorbic acid (C)	Scurvy
Choline	Fatty liver

Source: Ref. *54*.

(33) issued in 1973, calls for (1) acceptable evidence that the fortified food would be nutritionally and economically beneficial for a significant segment of the population; (2) that the food item selected would be a proper carrier for effective distribution of nutrient(s) to the population; and (3) that the added nutrient(s) would be physiologically available, chemically stable, capable of being monitored, and nontoxic in the food carrier under usual conditions of storage and dietary use.

Vitamins

About a dozen vitamins are available, by organic synthesis or by microbial fermentation, for fortifying foods.   Vitamins are essential for life and the well-being of man and animals, and biochemical functions of many are now known (Table 3.5).   Vitamins are classified separately from the mineral micronutrients because of their organic nature.   The limited range required, several micrograms to nearly 100 milligrams daily, separates them from the organic macronutrients, which are required in daily quantities of grams (proteins) to several hundred grams (carbohydrates and fats as sources of calories).

Minerals

In contrast to other nutrients, minerals are inorganic elements.   In nature, they usually occur in combination with other elements as "salts" or in organic compounds of varying complexity—for instance, iron in hemoglobin. While some have well-defined essential functions (Table 3.6), the role of others is less clear.   Essential functions have been demonstrated recently for very small intakes of certain trace minerals, such as selenium, that had been known previously only for their toxicity *(34)*.

Amino Acids

Amino acids are the fundamental building units of protein structure.   In the digestive process, proteins are broken down into their constituent amino acids, which the body can absorb.   Following absorption, provided the intake of carbohydrates and fats is adequate, the amino acids recombine to form the various proteins needed for body structure and function.   If the diet has insufficient calories (carbohydrates and fats), some of the amino acids will not be incorporated into protein but will be utilized to provide energy.   Similarly, excess amino acids, beyond those needed for protein synthesis, will be used to provide energy (calories) or converted to fat and stored in tissue.

TABLE 3.6   Physiological Functions of Minerals

Mineral (element in the form of a compound)	Physiological Functions
Calcium	Bone formation; various metabolic functions
Phosphorus	Bone formation; various metabolic functions
Magnesium	Bone constituent; carbohydrate and protein metabolism
Sodium	Water and acid–base balance; pressure; muscle function
Potassium	Acid–base balance; muscle function; carbohydrate and protein metabolism
Chlorine	Acid–base balance; gastric digestion
Sulfur	Various metabolic functions
Iron	Hemoglobin synthesis; cellular oxidations
Copper	Association with iron in enzyme systems and hemoglobin synthesis
Iodine	Synthesis of thyroxin, the thyroid hormone
Manganese	Activation of reactions in protein metabolism, glucose metabolism, and fatty acid synthesis
Cobalt	Constituent of vitamin B_{12}; essential factor in red blood cell formation
Zinc	Essential enzyme constituent; involved in storage of insulin
Molybdenum	Constituent of certain enzymes
Fluorine	Dental health
Selenium	Integrity of cell structures
Chromium	Glucose metabolism

Source: Refs. 55, 56

Most proteins are made up of combinations of some 20 different amino acids. Of these, eight—valine, leucine, isoleucine, threonine, methionine, phenylalanine, tryptophan, and lysine—are regarded as essential dietary compounds, with histidine possibly added for children and cystine for infants. Five of them—methionine, tryptophan, isoleucine, threonine, and lysine—may be consumed in insufficient quantities, particularly by children in some developing countries, depending upon the principal grain or root protein consumed.

Situations for Food Fortification

In addition to fortification, the dietary intake of a given population can be improved in a variety of methods, including educational efforts to increase nutrition awareness, increasing the quantity and/or improving the quality of food supplies, and distribution of food by governments. As a nutritional intervention method, fortification of foods can be especially effective in a situation where a substantial segment of a population would benefit from the incorporation of a nutrient or nutrients in its diet, but where the total amount of food available is reasonably adequate. The approach taken in a specific situation depends upon many factors: governmental policies; time period to achieve the goal; availability and suitability of a widely consumed carrier food for the nutrient; economics; expertise required; and safeguards to ensure that the efficacy and safety of the program will be maintained.   Situations that might call for fortification include:

- *Deficiencies in geochemical environments.*   Soils in wide areas of the world are deficient in certain minerals.   These situations can result in low concentrations of major or trace minerals in drinking water, plant crops, and even animal tissues, thus contributing to marginal or deficient dietary intakes in man *(34)*.   Fluoride is commonly added to water to reduce incidence of dental caries, and iodide to salt to control goiter.   In the future, other elements may also be used.

- *Inadequate supply or consumption.*   Some populations in the world do not have access to or do not consume enough foods of the needed nutritional quality.   While long-range approaches need to be introduced in such circumstances, a fortification program, if feasible, can introduce any missing nutrients more quickly.   An example is the insufficient consumption of iron in industrialized countries and vitamin A and iron in some of the developing countries.

- *Consumption of high calorie–low nutrient foods.*   With the passage of time, a greater percentage of the populations has become involved in more sedentary occupations and hence needs fewer calories.   On the other hand, advances in food technology have generated more kinds of food products, many of which are of a refined or highly processed character or may be fabricated from refined food components.   While many of these provide calories, they may not contain sufficient essential micronutrients; examples include the pure sugars, (sucrose, fructose, glucose), corn sugar syrups, table syrups, sweetened carbonated beverages, colored and sugared heat-processed beverages, vegetable shortenings, frying oils, salad oils and dressings, some snack foods, and even alcohol (ethanol).   The judicious fortification of some of the high-calorie foods, along with a nutrition education program to promote balanced consumption of other foods (including adequate protein), may be an effective approach to diet improvement.

- *Consumption of processed cereal products.*   About 26% of the daily caloric intake in the United States comes from products based on cereal grains; *(35)* this intake does not vary greatly with income and geographic region.   While adding variety and enjoyment to the daily menu, many processed cereal grain products have suffered substantial loss of vitamin and mineral content from the original whole grain in the manufacturing process.   Because of their wide usage and physical characteristics, cereal-grain products are suitable carriers for the addition of six vitamins and four minerals, according to the Food and Nutrition Board *(35)*.

- *Meal or food replacers.* Dietetic products (instant, frozen, or dried) have been developed and promoted to replace a regular meal. A suitable approach to fortification would be to have the new product provide a quarter or a third of the daily requirements of micronutrients. Likewise, when a new fabricated food product is developed to replace a natural food, the product should contain a reasonably complete and similar array of micronutrients.

Outlook

Knowledge of the requirements for and function of the essential micronutrients is not complete. The larger gaps in our knowldege of the major minerals are being filled in, and in the future, more trace minerals may be included in fortification programs. The addition of zinc has been proposed *(34)*. A better understanding of the utility of adding pure amino acids to food may lead to greater use in fortifying foods, provided costs are reasonable. New micronutrients can be incorporated only after detailed assessment of their normal intake, the human requirement, the nutritional status of the population, and, finally, possible use of other intervention measures. Appropriate labeling of fortified foods is also desirable to protect the small segment of the population that may be sensitive to increased intakes of certain nutrients—iron, for example. Further study is needed to determine how to protect these individuals.

The possibility that more micronutrients will be incorporated into fortification programs brings with it:

- Continued pressures on the chemist and food technologists to develop application forms that allow the nutrient to be uniformly incorporated and stabilized during the market distribution without altering flavor and biological availability in the human body. Technology has been worked out for adding vitamin A to sugar in Guatemala and to monosodium glutamate in the Philippines. In India and Indonesia, field trials are underway on fortification of salt with iron. Better technology is needed for addition of the various forms of iron, or of combinations of micronutrients, to certain foods.
- Continued evaluation of potential interactions among micronutrients that might interfere with performance in the body.
- Continued search for foods to fortify while minimizing changes in food habits and hence, reaching an increasing percentage of the population.

The economical production of pure and concentrated forms of the micronutrients essential for life has enabled nutritionists and food technologists to develop a nutrient delivery system that depends on the addition of missing or inadequate nutrients to existing foods or food ingredients. This method for improving the nutritional and health status of a population is a valuable tool in public health programs. Fortification, however, is certainly not a panacea. Its further application depends, in part, on new scientific advances in the understanding of human requirements; continued dietary, biochemical, and clinical surveys and surveillance; more data on nutrient content of foods; improved fortification technologies; and a better understanding of nutrient interrelationships.

FABRICATED FOODS

Fabricated foods are made by combining natural or synthetic components, or both, to achieve certain nutritional, sensory, and stability characteristics. (Traditional "formulated" foods such as bread, sausage, and wine will not be considered here.) Fabricated foods have been with us for some time, margarine and certain processed cheese foods being good examples. Recently, the emphasis has shifted to:

- Using more components partitioned from natural sources (oil or starch), or synthetic materials (amino acids or vitamins).
- Building in specific nutritional characteristics. For example, some foods for space flights were compounded on the basis of special needs (such as to supply additional calcium and potassium) under flight conditions.

Fabricated foods will generally be accepted only when the consumer is satisfied with their taste, odor, texture, and appearance. If fabricated foods meet this test, they have many potential benefits (Table 3.7). However, food materials are complex, the physical properties of food are poorly understood, and thus sensory characteristics can be difficult to duplicate.

In the industrialized countries, as well as in a number of the more advanced of the developing countries, societal pressures have led to changes in traditional methods of food production and consumption. These pressures include the need to reduce labor in the home, ensure nutritional quality and safety of foods, protect the environment, and make maximum use of limited resources. Fabricated foods are a response to some of these pressures. First of all, they provide flexibility. They can be made from a variety of materials and in combinations that are essentially equivalent nutritionally, making it possible to substitute one material for another during periods of shortages of specific raw materials. For instance, the proteins in milk are obtainable only from lactating animals. Fabricated high-protein

TABLE 3.7   Some Potential Benefits of Fabricated Foods

Provide complete infant foods
Provide "complete" diets for special uses
Modify caloric density
Specify ingredients and nutrients
Improve dietary patterns
Avoid disadvantages of some foods
Make better use of protein sources
Utilize by-products
Eliminate naturally occurring toxicants or deleterious compounds
Provide nutritious and time-saving convenience foods
Ensure uniform quality, palatability, and stability
Provide nutritious snacks that satisfy without excess calories
Supply some needs of developing nations
Stretch world food supplies
Supply good nutritive value at lower cost

Source: Ref. 57.

beverages such as milk analogs (substitutes) can utilize properly processed proteins from oil-seeds, grains, yeast, or even milk protein that has been extracted, processed, and stored during years of plenty to provide a reserve for years of shortages.   Another advantage of fabricated foods is that they can be produced with a consistent composition.   Furthermore, nutritionally balanced foods can be produced to meet the needs of certain population groups (such as infants or the aged) and individuals with specific metabolic problems (such as disorders of the liver or kidney) or specific diseases (such as hypertension, diabetes, or lactose intolerance).   Fabricated foods are especially useful in hospital feedings.   However, as it is difficult to exactly duplicate a food as it occurs naturally, there is the risk that a fabricated food will replace a superior natural product.   Therefore, as the use of fabricated foods increases, the risks and benefits will need to be compared to those of other food alternatives.

Types of Fabricated Foods.

Fabricated foods can be divided into four groups *(36)*:

- Special dietary foods that constitute the entire diet such as infant feeding formulas and diets for special uses such as control of weight or treatment of metabolic disorders (Table 3.8).
- Foods that can replace entire meals such as nutritionally complete beverages.

- Foods such as margarine, imitation fruit juices, or egg substitutes that simulate traditional foods.
- Minor foods such as imitation caviar, whipped toppings, and certain snack foods that ordinarily provide relatively few calories in the diet.

Major Research Needs

The future development of fabricated foods depends on increasing our understanding of the physical properties of foods.   The chemical industry developed its technology by characterizing the physical and physiochemical properties of the materials being processed, and then designing engineering processes based on those properties.   In contrast, the food industry has developed on a traditional basis, relying mainly on skilled artisans who passed qualitative and subjective procedures from generation to generation. Research is needed to develop methodology to measure and characterize the chemical, physical, and functional properties of food materials and to determine their role in the overall quality, processing suitability, and shelf stability of foods.

Another problem is the microstructure of foods, which is complex and heterogeneous.   Even a droplet of reconstituted spray dried milk differs markedly from the "homogenized" milk from which it is prepared.   Research is needed to relate the microstructure of both natural and fabricated food materials to their functional properties and processing characteristics. This research should include such areas as the microstructure of emulsions, finished products like bread, and texturized food such as extruded soybean products.   One of the goals should be to relate the microstructure to mech-

TABLE 3.8   Fabricated Foods Comprising Sole Diets

Foods for infants	Normal
	Allergic
	Gastrointestinal disorders
	Inborn errors of metabolism
Diets for special uses	Weight control
	Tube feedings
	Hyperalimentation
	Elemental diets
	Lactose intolerance
	Fat malabsorption
	Renal disease
	Cancer
	Metabolic studies

anisms of physical and chemical interaction within foods.   New structure-forming materials need to be studied. For instance, complex carbohydrates are still under-explored compared to proteins.

Nutritional aspects of fabricated foods are another area in need of a major research effort.   The most significant nutritional problems can be divided into two areas:

- *Nutritional adequacy,* which becomes important when a fabricated food constitutes a significant part of the total daily intake, or when a dominant contributor of a particular nutrient (for example, citrus juice as a source of vitamin C, certain B vitamins, and some minerals) is replaced by another food.   Thus, a meat substitute consumed as a snack food in addition to the major diet need not have the same standards of equivalence to meat as an entree product that accounts for 50% of total animal protein intake.   Another concern is the possible absence of nutrients for which requirements have not been established.   These include both known nutrients and others as yet undefined.   For example, recent information suggests an essential role for several trace elements, including silicon, chromium, and boron, not previously considered important in human metabolism *(37).*   Finally, if a product is a major source of a nutrient, its capacity to meet the nutritional needs of a variety of age groups and under a variety of health conditions becomes significant.

- *Nutritional imbalance,* which must be considered when a food or new composition is introduced, not in terms of interaction within the food but rather as a result of the changing overall daily nutritional intake.

The safety problem of fabricated foods, as of traditional foods, is far from simple.   New products must not only perform as claimed, but they must do so while providing the desired nutritional, sensory, and stability characteristics, with minimum use of additives.   A major problem in making such an assessment is associated again with age and physiological state of the consumer, since data are not available on the effect of many additives in young or old people and those under stress conditions *(38).*   In the case of fabricated foods it is possible to control additives and modify formulations during processing.   The important point is that modern technology, in causing us to reevaluate how we use and regard our food supply, requires a thorough evaluation of the associated biological problems *(39, 40).*

Outlook

A survey conducted in 1973 projected a total U.S. market for fabricated foods of $23 billion in 1980, compared to an estimated $12 billion in 1972 *(41).*   A large increase was projected for vegetable protein products, which were expected to reach sales of $1.5 billion, compared to the 1972 level of $80 million. Another forecast surveyed the opinions of experts as to the expected availability of nonconventional raw materials for fabricated foods *(42)*. Most rapid advances were expected in carbohydrates from cellulose and in utilization of proteins from plant materials. Single cell protein and synthetic materials for direct human consumption were considered less likely to be utilized by the food industry in the present century; however, others *(43)* consider these potential sources of food to be important areas for long-range research.

SPECIAL DIETARY FOODS

Advances in nutritional science and medicine have enabled nutritionists to formulate foods and diets to satisfy the special needs of selected population groups. The design of foods having modified fatty acid patterns and low cholesterol for individuals at risk of developing coronary artery disease is one example. Another is the special diets developed for those with inborn errors of amino acid metabolism. Diets that restrict one or more amino acids to the minimum essential for normal growth have been developed for phenylketonuria and a number of similar disorders *(44)*. In phenylketonuria, the body can break down only limited amounts of phenylalanine, an essential amino acid. The amount provided by the diet must meet the growth requirement, yet its intake must not permit an excess accumulation in body fluids of the amino acid or its degradation products. Two general kinds of dietary products are available for those suffering from phenlyketonuria. Products low in phenylalanine are made from suitably processed proteins, or from amino acid mixtures, and are fortified with the required additional nutrients to meet the standards of an infant formula. Phenylalanine-free products have an advantage for older patients, because they permit greater latitude in choosing natural foods containing phenylalanine to meet the dietary requirement for this essential amino acid. Such products also are made from proteins or are mixtures of amino acids.

Among the most significant problems requiring special dietary treatment are weight control, dental caries, and hypertension. For these problems, control over caloric intake and reduction in salt consumption are essential. Because ''sweet'' and ''salty'' are basic taste sensations, it is desirable to have nonnutritive sweeteners and salt substitutes to replace the natural sources of these sensations. Both illustrate the problems associated with development of foods for special dietary purposes.

Nonnutritive Sweeteners

The use of nonnutritive sweeteners is predominantly responsive to two popular demands:

- The desire to reduce caloric intake by replacing sugar in the diet.
- The need for a substitute for sugar in the diet of the diabetic individual.

Over two-thirds of the consumption of nonnutritive sweeteners has been in soft drinks; other products include canned fruits, vegetables, pickles, jams, jellies, and various sauces and dressings *(45)*.

Aside from dietetic purposes, several technological advantages have been cited for nonnutritive sweeteners:

- The lower sugar content of the liquid in canned fruit results in a plumper product because the osmotic effect carries water into the fruit rather than from the fruit to the syrup.
- The browning effect encountered with storage of sugar-containing products is reduced.
- Caramelization in cooking is eliminated.
- Reduction of sugar bulk reduces processors' handling costs.
- Certain naturally fermented pickles can be sweetened without sugar because noncaloric sweeteners will not ferment.

In October 1969, the FDA banned cyclamates because bladder cancer appeared in rats fed diets containing a sweetener made up of ten parts of cyclamate to one part of saccharin. At that time, annual consumption was estimated at about 21 million pounds of cyclamate and 4 million pounds of saccharin. With the withdrawal of cyclamate, saccharin use was estimated to have risen to 5 million pounds per year early in 1977. In April 1977, the FDA proposed banning the use of saccharin, again on the basis of bladder cancer in rats fed massive doses. In response to unprecedented public reaction—FDA received more than 100,000 comments, most opposing the ban—Congress, in November 1977, passed an 18-month moratorium on the ban. In addition, Congress asked the National Academy of Sciences to form a committee to assess not only the health risks and benefits of saccharin but to study several aspects of current food safety policy. The Committee for a Study of Saccharin and Food Safety Policy has issued two reports *(46, 47)* in which it recommends that FDA be allowed to assign food additives and contaminants into high, moderate, or low risk categories, and to consider both risks and benefits, particularly health benefits, in regulatory activities. Under that policy, saccharin would be in either the high or moderate risk category, and FDA would have a number of regulatory approaches available. The reports of the committee are likely to serve as a

TABLE 3.9   Nonnutritive Sweeteners

Common Name	Chemical Description	Sweetness (Sucrose=1)	Restrictions in Use	Comments
Saccharin	2, 3-Dihydro-3-oxobenziso-sulfonazole	300–500	Soluble in water as sodium salt; bitter aftertaste	Suspected carcinogen, banned by FDA but use extended to 1979 by Congress
Cyclamate	Cyclohexylsulfamic acid (sodium and calcium salts)	30–60	Bitter aftertaste at high concentrations	Banned by FDA in 1969
Aspartame	L-Aspartyl-L-phenylalanine methyl ester	100–200	Hydrolyzes in acid but may be encapsulated for soft drinks	Not yet approved for use
Chalcones	Flavones		Sweetness lingers and persists	Not yet approved for use
	Naringin	approx. 8000		
	Neohesperidin	approx. 160		
	Prunin	50–2000		
Terpenoids		200	May be bitter	Not yet approved for use
Oxathiazinone dioxide	Several derivatives		Not yet defined	Not yet approved for use
Kynurenine	Metabolite of tryptophane used as formyl and acetyl derivative	35	Not yet tested	Not yet approved for use
Monellin	Protein derived from serendipity berry *Discorephyllum cummunsii* Diels	3000	Lingering taste	Not yet approved for use

useful starting point for a national and Congressional debate on food safety. Under current law, FDA must propose the ban again after the moratorium expires.

The search for suitable substitutes for cyclamate and saccharin has intensified, and increasing knowledge of the chemical structures required for sweetness will assist in the search *(48)*. The several substitutes under investigation can be rated in sweetness relative to that of sucrose (Table 3.9). These relative values vary with conditions of testing, even with experienced tasters, and also with the kind or product being sweetened. In addition to taste qualities, various chemical and physical properties, such as stability to heat and solubility, are limiting factors.

Salt Substitutes

The development of diuretics to aid the body in removal of excess fluids has been a significant advance in the treatment of hypertension. However, many epidemiologists recommend lowering the intake of sodium salts as a preventive measure for the American population as a whole, and this recommendation was one of six made in "Dietary Goals for the United States," a report prepared for the U.S. Senate *(49)*. Some scientists disagree with the recommendation, believing that lowering intake of sodium salts for the entire population is not necessary *(50)*. However, it might be desirable for approximately 20% of U.S. children who, either from heredity or the environment, are at risk of developing hypertension as adults *(51)*. Although epidemiological studies have demonstrated an association between sodium salts intake and hypertension, it is as yet uncertain whether salt in the diet in fact induces hypertension. For the 20% of the population at risk, low salt intake begun early in life may protect, to some extent, against development of hypertension. Hypertensives who have been diagnosed are often prescribed a sodium-restricted diet. As defined by the American Heart Association, a mildly restricted sodium diet is limited to 3,000 mg of sodium (equivalent to 7.56 grams of salt, or sodium chloride) daily; a restricted diet is confined to 1,800 mg of sodium daily, and a severely restricted diet to 500 mg daily. It is extremely difficult for a person to meet a 500 mg per day dietary restriction. In many sections of the country, water used in cooking may provide 80% of this amount.

Many common foods today, particularly processed meats, contain substantial quantities of sodium salts. Even limited processing such as freezing and canning alters the sodium contents of fresh vegetables from the natural state *(52)*. For these reasons, people exposed to these products for virtually an entire lifetime tend to demand this amount of salt as a criterion for acceptability.

Because the psychological effect of "salting the food" is often perceived as a necessary ritual to increasing taste acceptability, a number of salt substitutes have been developed for use by persons on a sodium-restricted diet.   These are generally based on potassium chloride or mixtures of potassium chloride with ammonium chloride or sodium chloride.   Other ingredients such as glutamic acid and its salts are added to accentuate flavor and reduce the amount needed to intensify flavor.   However, they do not completely duplicate the flavor effects of sodium chloride.

REFERENCES, CHAPTER 3

1. "Chemicals Used in Food Processing," Food Protection Committee, Food and Nutrition Board, National Research Council; National Academy of Sciences: Washington, DC, 1965; publication 1274, pp. 31, 266.
2. "Toxicants Occurring Naturally in Foods," 2nd ed.; Report of the Committee on Food Protection, Food and Nutrition Board, National Research Council; National Academy of Sciences: Washington, DC, 1973.
3. Angeline, J.F.; Leonardos, G.P. "Food Additives—Some Economic Considerations," *Food. Technol.* **1973**, *27*(4), 40.
4. "The Risk/Benefit Concept As Applied to Food," A Scientific Summary of the Institute of Food Technologists' Expert Panel on Food and Safety and Nutrition, Institute of Food Technologists, Chicago, Illinois, March 1978.
5. Coggins, C.W.; Hield, H.Z.; Boswell, S.B. "The Influence of Potassium Gibberellate on 'Lisbon' Lemon Trees and Fruit," *Proc. Amer. Soc. Hortic. Sci.* **1960**, *76*,199.
6. Harvey, J.M.; Pentzer, W.T. "Market Diseases of Grapes and Other Small Fruits," *U.S. Dep. Agric., Agric. Handb.* **1960**, *189*, 1.
7. Salunkhe, D.K., et al. "On Storage of Fruits: Effects of Pre- and Postharvest Treatments," *Food Technol.* **1962**, *16*(11), 119.
8. Zink, F.W. "N^6-Benzlyladenine, a Senescence Inhibitor for Green Vegetables," *J. Agric. Food Chem.* **1961**, *9*, 304.
9. Fuller, Glenn, et al. "Use of Natural Cytokinins to Extend the Storage Life of Broccoli (*Brassica oleracea*, Italica group)," *J. Amer. Soc. Hortic. Sci.* **1977**, *102*(4), 480–484.
10. "The Use of Chemicals in Food Production, Processing, Storage and Distribution," Committee on Food Protection, Food and Nutrition Board, National Research Council; National Academy of Sciences: Washington, DC, 1973; p. 9.
11. Pintauro, N.D. "Food Additives to Extend Shelf Life"; Noyes Data Corp.: Park Ridge, NJ, 1974; p. 350.
12. Furia, T.E. "Handbook of Food Additives," 2nd ed.; CRC Press: Cleveland, 1972; pp. 44, 182, 229, 244, 277, 282, 598, 668.
13. Hofmann, K. "The Influence of Method of Preservation on the Quality of Meat and Meat Products," *Fleischwirtschaft* **1972**, *52*, 403.

14. "Accumulation of Nitrate," Committee on Nitrate Accumulation, Agriculture Board, Division of Biology and Agriculture, National Research Council; National Academy of Sciences: Washington, DC, 1972.

15. Pfeil, E.; Liepe, H.W. "Nitrosamine Problem," *Fleischwirtschaft* **1973**, *53*, 387.

16. "Nitrates: An Environmental Assessment," Panel of Nitrates of the Coordinating Committee for Scientific and Technical Assessment of Environmental Pollutants, National Research Council; National Academy of Sciences: Washington, DC, 1978.

17. "Scientific and Technical Assessment Report on Nitrosamines," U.S. Environmental Protection Agency, Office of Research and Development: Washington, DC, 1977; No. EPA 600/6-77-001.

18. "Statement on Nitrites," Press release by the Food and Drug Administration and U.S. Department of Agriculture, Washington, DC, August 11, 1978.

19. Harris, R.S.; Karmas, E. "Nutritional Evaluation of Food Processing"; AVI Publishing Co.: Westport, CN, 1975; pp. 262, 276, 337, 358.

20. Tressler, D.K.; Van Arsdel, W.B.; Copley, M.J. "The Freezing Preservation of Foods"; AVI Publishing Co.: Westport, CN, 1968; p. 316.

21. Salunkhe, D.K.; Do, J.Y.; Bolin, H.R. "Developments in Technology and Nutritive Value of Dehydrated Fruits, Vegetables, and Their Products," *Crit. Rev. Food Technol.* **1973**, *4*(2), 156.

22. Bolin, H.R.; Boyle, F.P. "Use of Potassium Sorbate, DEPC, and Heat for the Preservation of Prunes of High Moisture Levels," *J. Sci. Food Agric.* **1967**, *18*, 289.

23. Schroeter, L.C. "Sulfur Dioxide Applications in Foods, Beverages, and Pharmaceuticals"; Pergamon: New York, 1966; p. 204.

24. Powers, J.J. "Effect of Acidification of Canned Tomatoes on Quality and Shelf Life," *Crit. Rev. Food Sci. Nutri.* **1976,** *7*(4), 393.

25. National Food Guide, U.S. Dept. Agric., ARS, Inst. Home Econ., Washington, DC, August 1946, Leaflet No. 288 (formerly AIS-53 and NFC-4).

26. Food for Fitness, A Daily Food Guide; U.S. Dept. Agric., ARS, Inst. Home Econ., Washington, DC, March 1958, Leaflet No. 424.

27. Recommended Dietary Allowances, 9th ed.; Food and Nutrition Board, National Research Council; National Academy of Sciences: Washington, DC, 1979.

28. Bauernfeind, J.C. Vitamin Fortification and Nutrified Foods. *Proc. Int. Congr. Food Sci. Technol.* **1970**, *3*, 217–232.

29. LaChance, P.A. "Nutrification, A New Nutrition Concept for New Types of Foods," *Food Technol.* **1970**, *24*(6), 724.

30. "Food Fortification: Protein-Calorie Malnutrition," *W.H.O. Tech. Rep. Ser.* **1971**, *477*.

31. Boussingault, M. "Memoir sur les Salines Iodiferes des Andes," *Ann. Chem. Phys.* **1833**, *54*, 163–177.

32. Morton, R.A. "The Vitaminization of Margarine," *J., R. Soc. Health* **1970**, *90*(11), 21–28.

33. General Policies in Regard to Improvement of Nutritive Quality of Foods. Food and Nutrition Board, National Research Council; National Academy of Sciences: 1973; Vol. 31, Number 10, pp. 324–326.

34. Mertz, W. "Fortification of Foods with Vitamins and Minerals," *Ann. N.Y. Acad. Sci.* **1977**, *300*, 151–160.

35. Proposed Fortification Policy for Cereal-Grain Products. Food and Nutrition Board; National Research Council–National Academy of Sciences: Washington, DC, 1974; publication ISBN 0-309-02232-0.

36. Sarrett, H.P. "Potentials and Hazards of Engineered Foods"; lecture presented at the Johns Hopkins University International Symposium on Nutrition and Public Health, 1975.

37. "Recommended Dietary Allowances," 9th ed.; Food and Nutrition Board, National Research Council; National Academy of Sciences: Washington, DC, 1979.

38. Sui, R.G.H., et al. "Evaluation of Health Aspects of GRAS Food Ingredients: Lessons Learned and Questions Unanswered." (Report of SCOGS, LSRO, FASEB) *Fed. Proc., Fed. Am. Soc. Exp. Biol.* **1977**, *36*, 2519–2562.

39. Miller, S.A. "Food and Nutrition in Health and Disease," *Ann. N.Y. Acad. Sci.* **1978**, *300*, 397–405.

40. Miller, S.A. "Risk/Benefit, No-Effect Levels and Delaney: Is The Message Getting Through?" *Food Technol.* **1978**, *32*, 93–96.

41. "Fabricated Food Market To Exceed $23 Billion by 1980," *Food Technol.* **1973**, *27*(12), 46.

42. Holmes, A.W. "Substitute Foods: A Practical Alternative?" *Philos. Trans. R. Soc. London, Ser. B* **1973**, *267*(882), 157.

43. "Feeding the Expanding World Population: International Actions to Avert the Impending Protein Crisis," United Nations, New York, 1968.

44. "Special Diets for Infants with Inborn Errors of Amino Acid Metabolism," *Pediatrics* **1976**, *57*(5), 783.

45. "Food Consumption, Pricing & Expenditures," Economic Research Service, U.S.D.A. Supplement for 1973 to Agriculture Economics Report No. 138. (Tables 38 & 39 which did refer to nutrients available to the percentage of nutrients available from meat diets.)

46. "Saccharin: Technical Assessment of Risks and Benefits," National Research Council/Institute of Medicine, Committee for a Study on Saccharin and Food Safety Policy. National Academy of Sciences: Washington, DC, 1978.

47. "Food Safety Policy: Scientific and Societal Considerations." National Research Council/Institute of Medicine, Committee for a Study on Saccharin and Food Safety Policy, National Technical Information Service: Springfield, VA, 1979; PB 292 069.

48. Shallenberger, R.S.; Acree, T.E. "Chemical Structure of Compounds and Their Sweet and Bitter Taste," In "Handbook of Sensory Physiology"; Springer Verlag; New York, 1971; Chapter 12, Part 2, pp. 221–227.

49. "Dietary Goals for the United States," (revised edition), Select Committee on Nutrition and Human Needs. U.S. Government Printing Office: Washington, DC, 1978.

50. "Dietary Goals for the United States: A Commentary," Council for Agricultural Science and Technology, Iowa State University, Ames, IA, 1977, No. 71.

51. "Sweet Intake and Eating Patterns of Infants and Children in Relation to Blood Pressures," *Pediatrics* **1974**, *53*, 115.

52. Bills, C.E.; McDonald, F.G.; Niedermeier, W.; Schwartz, M.C. "Sodium and Potassium Analysis of Foods and Waters," *J. Am. Diet. Assoc.* **1949**, *25*, 304–314.

53. "1978 Handbook of Agricultural Charts," *U.S. Dept. Agric., Agric. Handb.* **1978**, and earlier editions.

54. "Encyclopedia of Food Science"; Peterson, Martin S., Johnson, Arnold H., Eds.; AVI Publishing Co., Inc.: Westport, CN, 1978.

55. Amen, R.J. *Food Prod. Dev.* **1973**, *7*(9), 33–34.

56. Ibid., **1973**, *7*(10), 74–84.

57. Sarrett, H.P. "Potentials and Hazards of Engineered Foods," Johns Hopkins University International Symposium on Nutrition and Public Health, 1975.

58. Durrenmatt, Konrad. "Part Five Interpretive Summary: Feasibility and Commercial Aspects of Amino Acid Fortification," In "Amino Acid Fortification of Protein Foods"; Scrimshaw, N.S., Altshul, Aaron M., Eds.; MIT Press: Cambridge, MA, 1971.

Unconventional Sources of Food and Feed

The world food and feed situation has encouraged a special effort to develop unconventional sources that would be nutritious, acceptable, and of moderate cost. In recent years, emphasis has centered on single-cell proteins and waste materials; there has also been some interest in synthetic fats and carbohydrates for use in highly processed fabricated foods. While new technologies have been developed to produce foods and feeds, the major drawback has been the high cost of the materials produced.

SINGLE-CELL PROTEIN

At the First International Conference on Single-Cell Protein held in 1967, the term "single-cell protein," or SCP, was used to describe microbial sources of protein (1). Although the term is new, SCP has long been consumed as an integral part of the food supply, although usually not in isolated form (Table 4.1). In modern times, various attempts have been made to use microbial cell mass or SCP as a major protein source. Stimulated by the difficulties of war and the availability of a raw material in the form of single sugars from wood hydrolysis, as much as 16,000 tons of yeast per year were incorporated into human food during World War II (1).

SCP has shown promise, although, so far, the use of the newer types has been mainly for animal feed. For a number of reasons, in part cultural and historical, yeast has received the greatest attention. Yeast protein has been a part of man's diet since the advent of alcoholic beverages, making it an acceptable ingredient for food and feed. Yeast protein is easily pro-

TABLE 4.1   Traditional Foods Containing SCP

Food	Microorganism(s)
Cheese	
Cheddar	*Streptococcus lactis, Lactobacillus casei,*
Roquefort	*Penicillium roqueforti*
Swiss	*Streptococcus lactis, Lactobacillus bulgaricus, Propionibacterium shermanii*
Other dairy products	
Sour cream	*Streptococcus* sp.
Yogurt	*Lactobacillus brevis, Leuconostoc mesenteroides*
Kieffer	*Lactobacillus* sp.
Sauerkraut	*Leuconostoc mesenteroides, Lactobacillus plantarum, Lactobacillus brevis*
Pickles	*Acetobacter* sp., *Pediococcus cerevisiae, Lactobacillus brevis, Streptococcus faecalis*
Soy sauce	*Asperigillus oryzae, Lactobacillus delbrueckii, Zygosaccharomyces soyae*
Mushrooms	*Agaricus bisporus*
Fermented sausages	*Lactobacillus* sp., *Pediococcus sp.*

Source: Refs. *18, 19.*

duced in large quantities and can utilize a wide range of nutrients.   With
further processing, it exhibits a variety of desirable sensory properties and is
readily incorporated into feed and food products.   The major constraint is
economics.   Another hindrance is the creation of medical problems: nucleic acids in yeast can increase blood uric acid levels in a segment of the
population, leading to an increase of gout.

Nutritional Requirements of SCP Production

The major elements composing microbial cells (carbon, oxygen, hydrogen,
and nitrogen) as well as the minor elements (phosphorus, magnesium,
calcium, sulfur, and potassium) and trace elements (cobalt, zinc, copper,
and molybdenum, for example) must be made available for cell growth.
Energy must be provided, and certain vitamins and amino acids may also be
required, depending on the specific microorganism being grown.
     The carbon source is the most variable.   Carbon in the form of glucose,
however, can be utilized by virtually all microorganisms.   Yeasts (chiefly
Candida and *Saccharomyces*), as well as some other microorganisms, can
utilize a range of carbon sources, from simple sugars, to starches, to paraf-
finic hydrocarbons.

TABLE 4.2   Carbon Sources of SCP in or Nearing Commercialization

Carbon Source	Organism	Company	Location
Crude oil fractions			
n-Paraffins	Candida	British Petroleum[a]	Grangemouth, Scotland
Gas oil	Candida	British Petroleum	Laveva, France
Gas oil	Candida	———	USSR
Methanol	Methylomonas methylotropha	ICI	England
Methanol	Methylomonas clara	Hoechst	Frankfurt, West Germany
Ethanol	Candida	Amoco Foods	Minnesota
Lactose	Saccharomyces	Milbrew	Wisconsin
Wood hydrolyzate	Saccharomyces	———	USSR
Sulfite waste liquor	Microfungus	SITU Group	Finland

[a] BP has recently decided to discontinue their SCP effort.

The primary criteria in the selection of the carbon source are ready availability at a low cost and freedom from undesirable contaminants. Various sources have been developed throughout the world depending upon local supplies and economics; a number are currently in pilot plant or commercial use (Table 4.2). Other sources well along in development were identified in a report to the National Science Foundation (2):

- Methane
- Cellulosic residues (bagasse, waste paper, hay, etc.)
- Carbon dioxide
- Carbohydrate-containing crops and residues (citrus waste, carob, cannery wastes, coffee pulp, banana waste, etc.)
- Starchy crops (cassava, potato waste, etc.)

Economics

The primary use of SCP is a supplement to animal feeds. Its main competitor in this market is oil-seed meal, especially soybean meal. To compete, SCP must cost about the same as soybean meal. Alternatively, the properties of SCP must be upgraded for entry into the human food market or special markets such as pet or fish foods, where it can command premium prices.

TABLE 4.3   Relative Protein Costs

Protein	$/lb Dry Protein[a]
Pork meat	5.00
Poultry	2.00
Egg albumin	1.50
Gelatin	0.70
Casein	0.53
SCP produced from petrochemicals	0.40–0.70
Soya isolate	0.40
Ground nut isolate	0.38
Soy bean meal	0.20–0.30
SCP recovered from brewing wastes	0.15–0.35
Alfalfa	0.15

Source: Ref. 3.

[a] Quality of proteins for human nutrition is variable and must be considered in comparing costs.

Production costs of SCP are strongly influenced by the raw material costs *(3)*. In most conventional submerged fermentation processes, 35 to 60% of the SCP production costs can be attributed to the raw material. SCP costs vary from 15 to 70 cents per pound of protein, putting it in a potentially competitive position (Table 4.3). Unfortunately, this favorable price is realized only in large plants producing approximately 100,000 tons per year and costing about $60 million (in 1978 dollars) *(4)*. One such plant would produce enough SCP to replace 5% of the soy meal now used in animal feeds. Thus there is little room left to reduce costs further by increasing plant size.

Nutritional Considerations

Thorough clinical testing of a number of different SCP's grown on a wide variety of carbon and energy sources has established that appropriately processed SCP is nutritious and safe as a dietary supplement for humans and animals. The nutritional value of various SCP's compares quite favorably with solvent-extracted soybean meal and fish meal. The crude protein level varies from 31 to 78%, compared to 45% for soy and 66% for fish meal. Except for the fungal protein, the essential amino acid profile also compares quite favorably with soy and fish meals.

The important nutritional limitations on the use of SCP are the relatively high levels of nucleic acid and low levels of two essential amino acids, methionine and tryptophan. The purine components of the nucleic acid in SCP are generally converted to uric acid in humans. Since uric acid is rather insoluble in body fluids, it can lead to medical problems such as gout and kidney stones.   The problem of elevated nucleic acid levels has led to the development of special mutant strains of temperature-sensitive micro-organisms that excrete their nucleic acid when the temperature shifts (2). (It is possible, however, to reduce the nucleic acid levels by an extraction process).   These special strains are probably being used in some of the commercial processes under development.

Most human foods contain nucleic acids at various levels.   The nucleic acid content of SCP varies from 8 to 25 grams per 100 grams of protein, while liver contains 4 grams and wheat flour, 1 gram (2).   Therefore, if SCP is to be used directly, an acceptable level of nucleic acids in the human diet must be determined.   Nucleic acid does not pose a problem to most farm animals because they can convert uric acid to a soluble compound (allantoin).

An additional problem concerns the level of the trace contaminants carried over from certain raw materials (such as aromatic hydrocarbons in gas oil), which are under intense scrutiny by regulatory officials.   Regulations in Italy, for example, have precluded economical production by one proposed commercial process.

Future Prospects

A number of factors combine to keep the potential of SCP as a *future*, major source of protein quite high.   Chief among these is the exponential growth of the world's population.   Of the potential alternate food sources, SCP holds promise for the following reasons (2):

- SCP production requires relatively little space—theoretically the world's food supply could be supplemented by 10% in an area of 1.5 square kilometers.
- SCP is not dependent on the agricultural sector for raw materials.
- SCP processes present minimum environmental problems.
- There is a long human experience in use of foods containing SCP.

Unless the price of extracted soy meal rises significantly, it will be difficult for an SCP process to be economically competitive as a crude protein source.   Some commercial ventures are attempting to market SCP for human use or for specialized animal feeds.   The most likely near-term use of SCP will continue to be as a component of animal feeds and in human foods, primarily as a processing aid to provide certain sensory properties.

WASTES

Production of proteins from wastes must consider both the nature of the waste and the use of the protein. The major consumer of protein in the developed world is the animal feed industry, followed by human consumption. A significant fraction of the protein in animal feeds consists of crude agricultural by-products that were considered wastes prior to their acceptance by the animal feed industry. Even soybean meal once had little value. Once these "wastes" gained a market, they ceased to be wastes and became a valuable by-product. Today, even agricultural residues with relatively low protein contents have a place in the animal feed market as long as the combination of protein and energy is suitable.

Conversion of agricultural wastes or residues to human foods is quite restricted. Questions of consumer acceptance and safety must be resolved before human consumption can increase. Technology must also be improved so that the proteins can be economically recovered, purified, and processed from wastes. On the other hand, recovery of whey solids from dairy operations for human food is feasible and alleviates environmental problems.

If a waste is to be useful as a source of protein for animal feeds, it must meet a number of criteria, including:

- Protein content must be high enough (or easily raised to a high enough level) to be incorporated into feeds.
- The waste must be free of any toxic materials.
- The supply must be large, reliable, and of uniform quality.
- The price must be competitive with alfalfa or oil seed on a protein basis.

The protein content can be raised by separating the protein from the waste or by fermenting a portion of the waste to single-cell protein. To be competitive, the protein upgrading step must cost less than 5 cents per pound. This cost is quite sensitive to the quantity processed, so that more than 100,000 tons per year of waste are needed for upgrading in a central processing plant. Unfortunately, few unutilized protein wastes are available in sufficient quantity to justify more than a minimal capital investment for any further processing.

Types of Waste

A large variety of solid, organic wastes are potential sources of crude protein (Table 4.4) *(5,6)*.

TABLE 4.4   Major U.S. Solid Wastes

Waste	Million Tons/Year
Agricultural	600
Large sources of animal waste	(26)
Manure	(200)
Large canneries, mills, slaughter houses, and dairies	(23)
Cereal straw	(161)
Cornstalk	(142)
Other	(48)
Urban and municipal solid wastes	160
Wood manufacturing	21
Logging and other wood wastes	60
Industrial wastes	45
Municipal sewage solids	15
Miscellaneous organic wastes	70
TOTAL	971

Source: Refs. 5, 6.

Most sources of agricultural wastes are very dispersed, creating a collection problem that adds significantly to the cost of producing crude protein. Also, many of the major waste categories have value as an energy source, soil conditioner, and/or fertilizer. In addition, many categories are either undefined or variable in composition and quality. This is particularly true of urban, industrial, and sewage wastes. Hence, it is almost impossible to guarantee the safety of these wastes as a source of food and feed. However, several large-tonnage wastes are potential sources of protein and/or energy, including cornstalk, cereal straw, and wood waste. The use of spent sulfite liquor from the wood pulp industry as the carbon source for single-cell protein is well established. In addition, technological advances continue to be made—for example, the PEKILO process for spent sulfite liquor, which is being developed in Finland (7). A single 100,000-ton-per-year pulp mill can produce 10,000–15,000 tons per year of microfungal cell mass with a crude protein content of 55–60%. The Finnish government is building a semicommercial plant to determine the economics of the process.

In recent studies of straw (8), a single semisolid fermentation using mixed cultures of microorganisms increased the crude protein content to 13.9%, a fourfold increase. Since this figure is close to the crude protein content of alfalfa, it is now possible to process straw to a material suitable for animal feeds.

Processes for converting a number of wastes to feed materials, high quality proteins, and energy (in the form of ethanol and methane) are being studied at the laboratory level. Among the wastes are: poultry feathers, green leaf protein, tobacco leaf protein, distillers by-products, whey, animal by-products (blood, skin, hair, horns, etc.), and chitin from shellfish.

Although the processes have succeeded in recovering and upgrading the protein in these wastes, the economics have been highly unfavorable for commercialization. The typical problems encountered have been: the process is very complex; the dispersion of waste is too great to justify a large central processing plant; and the yields of high-quality protein are too low.

Future Prospects

As the price of energy continues to climb faster than the price of agricultural commodities, the food and energy sectors of the world economy will increasingly compete for agricultural residues and land. This is illustrated by the cultivation of sugarcane, manioc, and other plants in Brazil solely for fermentation into ethanol (9). Plans call for producing 1.2 billion gallons of ethanol by 1986. Although the economics of producing ethanol from agricultural crops in the United States are currently unfavorable, the use of agricultural residues as a source of energy is already significant. Sugar cane bagasse is used extensively as a boiler feed, and major development efforts are underway for the anaerobic conversion of agricultural residues to methane.

This increasing competition for various organic wastes as energy sources will reduce the potential economic incentives for their recovery and/or conversion to proteins. Protein from fermentation wastes will continue to grow, however, and gain an increasing (although small) share of the animal feed market. Such growth will free larger quantities of oil seed meal for use in nutritious fabricated foods for human consumption.

SYNTHETIC FATS AND CARBOHYDRATES

The synthesis of traditional fats and carbohydrates by chemical means is one approach to increasing the supply of energy-supplying food. This approach has the advantage of freeing food production from the dependence on land and the vagaries of climate.

Synthetic Fats

Natural fats and oils consist of various glycerides of fatty acids, the distinction being that fats are solids at room temperatures, while oils are liquids. To synthesize fats or oils, the two major components, fatty acids and glycerol, are reacted.   A typical reaction is that involving oleic acid, an 18-carbon unsaturated acid:

$$3\ C_{17}H_{33}COOH\ +\ \begin{array}{c} CH_2OH \\ | \\ CHOH \\ | \\ CH_2OH \end{array} \quad \xrightarrow[catalyst]{heat} \quad \begin{array}{c} CH_2OCOC_{17}H_{33} \\ | \\ CHOCOC_{17}H_{33} \\ | \\ CH_2OCOC_{17}H_{33} \end{array}$$

The two most likely starting materials are hydrocarbons from petroleum and simple gases such as carbon monoxide (CO) and methane (CH_4), which are derived from petroleum.   Petroleum remains the most abundant raw material for large-scale food synthesis, although, due to the current energy crisis, its price will increase.

The simplest method of synthesizing fatty acid precursors is to separate the paraffin or "straight chain" hydrocarbons from petroleum.   The hydrocarbons can then be converted to fatty acids through oxidation and purification.   The Fischer–Tropsch reaction (10) developed in Germany in 1923, affords an alternate route to fatty acid precursors.   This reaction can convert carbon monoxide and hydrogen to straight-chained paraffins, which can then be converted to an edible product.   The Germans actually employed this process during the World War II to produce margarine.   The synthetic product varied considerably in quality, but was apparently nontoxic and somewhat nutritious.   However, the low overall yield of edible material makes the process unattractive.   Still, the Fischer–Tropsch synthesis employs readily available starting materials (carbon monoxide and methane).   This technology, old and exhaustively studied (11, 12, 13), is currently experiencing a resurgence of interest.   New developments should be watched for clues in learning how to modify the synthesis to produce more straight-chain hydrocarbons.

The third method for synthesis of fatty acid precursors involves the conversion of CO to ethylene, $CH_2{=}CH_2$. followed by controlled polymerization by means of the reaction discovered by Karl Ziegler in Germany in 1959.   This approach produces long-chain olefins (open-chain unsaturated hydrocarbons containing at least one double-bonded carbon) of the right size with virtually no branched chain or other by-products (11, 14).

Traditional methods of direct oxidation of paraffins to fatty acids yield a product that is difficult to purify.   More recent methods use oxidizing agents such as chromic and nitric acids.   The reliable method for oxidizing an olefin is with ozone, provided appropriate precautions are taken for certain hazardous parts of the process.   The major advantage is yields of edible fatty acids from 90 to 95%.   The esterification of fatty acids with glycerol presents no special problems.

NUTRITIONAL AND TOXICOLOGICAL CONSIDERATIONS What characteristics must a synthetic fat possess to be nutritious and gain consumer acceptance? One analysis *(11)* cited the following characteristics:

- The fatty acids should be mostly straight chains of 10–16 carbon atoms.
- Odd-number carbon acids should be held to a minimum. (Natural fats consist overwhelmingly of even-number carbon acids.) Although some data suggest that the feeding of odd-number carbon fatty acids does not produce deleterious effects, no long-term studies have been published in which these fatty acids were fed at high levels.
- Some unsaturation of these acids is desirable, but excessive amounts can cause rancidity problems.

The likely distribution of fatty acids obtained from the Ziegler reaction followed by oxidation via ozone seems satisfactory *(10, 11)*. An unanswered question is whether this route would provide only odd-number-carbon fatty acids. However, it can be modified to produce even-carbon acids, if necessary, although this would add to the costs.

Work is required to establish the maximum tolerable limits of by-products in fatty acid produced by oxidation of paraffins. Synthetic fats would need to be supplemented with small amounts of certain fatty acids essential to human nutrition.

Synthetic Carbohydrates

One approach to synthesis of carbohydrates is the conversion of formaldehyde to a complex mixture of high molecular weight carbohydrates. The mixtures can be purified by conventional means to yield a material called "formose sugars" *(15)*.

The testing of the nutritive value of formose sugars is still at the level of animal experimentation *(16)*. Studies indicate that rats can tolerate a diet in which up to 25% of natural sugars is replaced with synthetic formose. Higher levels, however, are not well tolerated. The toxic effect of some formose preparations may be due to the presence of branched sugars not found in nature. If these could be completely removed, a nutritionally acceptable formose product seems feasible.

The formose reaction can also provide a route to glycerol, an energy source that the body can metabolize. In a 90-day study of glycerol as a human diet supplement, no nausea or other ill-effects were observed, although the glycerol content of the subjects' urine rose *(17)*. The synthetic formose sugars could be produced from formaldehyde, then processed to produce a mixture of glycerol and by-products, from which the glycerol could be separated.

Outlook for Synthetic Calorie Sources

The technical feasibility of chemical synthesis of fats seems well established. However, studies are required on their toxicology and nutritional value. Once the biological guidelines as to what types and amounts of impurities can be tolerated are established, costs can be determined. Consumer acceptance does not appear to present any serious problem.

There seems little likelihood that formose sugars will ever become a significant component of the human diet. Therefore, alternate routes will have to be found if synthetic carbohydrates are to contribute to future food supplies.

REFERENCES, CHAPTER 4

1. "Single-Cell Protein," Mateles, R.I., Tannenbaum, S.R., Eds.; MIT Press: Cambridge, 1968.
2. "Protein Resources and Technology: Status and Research Needs," Milner, M., Scrimshaw, N., Wang, D., Eds.; AVI Publishing Co.: Westport, CN, 1978
3. Moo-Young, M. "Economics of SCP Production," *Process Biochem.* May 6–10, **1977**.
4. Humphrey, E. "Current Developments in Fermentation," *Chem. Eng. (London)* Dec. 9, **1974,** p. 98–112.
5. Humphrey, E. "Current Developments in Fermentation," *Chem. Eng. (London)* Dec. 9, **1974,** p. 98.
6. Burwell, C.C. "Solar Biomass Energy: An Overview of U.S. Potential," *Science* **1978,** *199,* 1041.
7. "Protein from Spent Sulfite Liquor," *Process Biochem.* November **1973,** p. 19.
8. Anderson, A.W.; Han Y.W. "Increasing the Nutritive Value of Straw by Microbial Growth in a Semi-Solid Straw Matrix," paper presented at the conference on Mechanisms and Kinetics of Uptake and Utilization of Substrates for Single-Cell Protein Production, Moscow, June 1977.
9. "Biomass: The Great Green Hope," *Chem. Eng. News,* July 24, **1978,** p. 31.
10. "Synthetic Fats, Their Potential Contribution to World Food Requirements," The Division of Nutrition of the FAO, Nutritional Studies Series **1949,** publication No. 1.
11. "Study of Methods for Chemical Synthesis of Edible Fatty Acids and Lipids," Frankenfield, J.W., Ed.; Linden, NJ, 1967, Final Technical Report, NASA Contract NAS-2-3708. Exxon Research and Engineering Co.
12. Anderson, H.C.; Wiley, J.L.; Newell, A. U.S., *Bur. Mines Bull.* No. 544, Part I: Literature, 1954, and Part II: Patents, 1955. (A review of Fischer-Tropsch with 8,000 references.) **1954,** *544,* part I.
13. Ibid., **1955,** ———, Part II.

14. Ziegler, K. "The History of an Invention," *Angew. Chem.* **1965**, *76*, 545.
15. Shapira, J. "Biochemical Methods for the Synthesis of Potential Foods," *J. Agric. Food Chem.* **1970**, *18*, 992.
16. Partridge, R.D.; Weiss, A.H.; Todd, D. *Carbohydr. Res.* **1972**, *24*, 29–44.
17. Shapira, J. "Use of Glycerol as a Diet Supplement During a Ninety-Day Manned Test," paper presented at Symposium on Preliminary Results from an Operational 90-Day Manned Test of a Regenerative Life Support System, Hampton, VA, November 1970, NASA SP-261.
18. Prescott, S.C.; Dunn, C.G. "Industrial Microbiology," 3rd ed.; McGraw–Hill: New York, 1959.
19. Litchfield, J.H. "Microbial Cells on Your Menu," *Chem. Technol.* **1978**, *218*.

Assuring the Wise Use of Chemicals

Chemistry has been intimately—although not always obviously—involved in the food system throughout history. The primitive farmer recognized the value of fertilizers and pesticides, though he was confined to those readily available in nature. Similarly, people have long recognized the value of using additives (usually in the form of crude natural products or homemade mixtures) in their own foods to preserve them or to produce other desirable effects, such as improved flavor. However, the problem has been, and continues to be, to assure that chemicals are used wisely in the food system. Careless or excess use of pesticides can damage public health and the environment. Spices, while helping to preserve food and provide certain flavors, also provide the opportunity for covering up the flavor of spoiled foods. Additionally, some of the chemicals in the smoke used in food processing may be carcinogenic.

As foods became a matter of commerce, the producers used various materials in the processing of food for others. Avarice dictated that many such uses be to the economic disadvantage of the buyer. In his poem "Maud" (1855), Alfred Lord Tennyson commented on a flagrant example of adulteration: "Chalk and alum and plaster are sold to the poor for bread, and the spirit of murder works in the very means of life." In addition to avarice, ignorance and indifference sometimes resulted in harm to the consumer. For example, at one time formaldehyde, a strong irritant, was used to preserve milk, and lead salts were used to clarify wines.

Concern with both adulteration and the safety of food additives has been evident at least since the writings of Frederick Accum in the early 19th century. An abrasive and vitriolic social activist, Accum was the first to apply the new science of chemistry to the analysis of food for the protection of the consumer.   His publication "A Treatise on Adulteration of Food and Culinary Poisons" (1820) (*1*) generated bitter opposition, but led ultimately to the British food laws of 1860 and 1875.

The first U.S. law, the Food and Drug Act of 1906, followed that of other countries, prohibiting "any poisonous or deleterious substance" in food and defining as adulterated any food containing such a substance.   The Federal Food, Drug, and Cosmetic Act of 1938, which replaced the 1906 Act, contained substantially the same wording (*2*).   This law implied, but did not require, safety evaluation, leaving to the government the need to prove that harm could or did occur.   However, as chemical intervention in the production and processing of food increased, it became apparent that this post hoc approach to safety was clearly inadequate.   It was replaced by a principle embodied in three major amendments to the 1938 Act—the Pesticide Amendment of 1954 (*3*) the Food Additives Amendment of 1958 (*4*) and the Color Additives Amendments of 1960 (*5*). Under this principle, *the safety of a substance must be evaluated prior to its use and at the expense of the producer or industrial user.* Although not directly relevant to foods, recent drug amendments (*6*) and the Toxic Substances Control Act (*7*) demonstrate the same concerns by extending the concept of pretesting and preclearance to other uses of chemicals.

EVALUATION OF SAFETY

Safety evaluation is not a one-time activity but must occur both prior to and continually during use.   Ideally, evaluation of safety prior to use should recognize and intercept significant risks, but not be so burdensome as to unduly discourage innovation.   Continuing evaluation requires reexamination, as better techniques are developed for gathering, analyzing, and interpreting data on the human and environmental impacts resulting from use of chemicals in the food system.

Implicit in any evaluation of safety are two essential concepts:

- The principle of commensurate effort (the extent and cost of evaluation should be scaled to the probable risk).
- Recognition of the inevitable incompleteness of any evaluation, which stems from imperfect knowledge, lack of adequate techniques to measure and evaluate benefits and risks, and the impracticality of performing every theoretically conceivable test or examination.

The principle of commensurate effort is easier to state than to apply. Absolute safety means the absence of all risk, an unachievable state. Risk, which is the probability that a given use of a particular substance will result in harm, is a composite of several factors, chief among them being extent of exposure, degree of toxicity, and susceptibility of the organism exposed. If exposure can be measured or estimated with confidence, that aspect of probable risk can be partially evaluated in advance. Toxicity and susceptibility, however, are less easily dealt with a priori.

Most biological experiments, including toxicological tests and metabolic studies, reveal additional questions that need answering. The inherent complexity and variability of living organisms, and the inability to control every experimental condition inevitably cause some measurements to depart from control values. The proper interpretation and understanding of such aberrations usually require further testing. Thus the "definitive test" usually seems to be the one planned, never the one just completed.

It is easy to agree that in evaluating the possible level of risk of a new chemical additive or pesticide, one should err on the side of too much information rather than too little. The question is always: "How much more?" The search for more information on any one point inevitably diverts resources from other efforts. National and international testing capabilities and expertise, though quite extensive, are necessarily limited and should be devoted to solving problems most pertinent to human safety.

Under the 1954–60 amendments to the Food, Drug, and Cosmetic Act, the burden of establishing the safety of a substance usually falls on the manufacturer. The most heavily used additives have sufficient present or potential use to justify almost any reasonable testing program. The cost of testing additives used in smaller volume, such as many of the food colors, usually can be borne only if several manufacturers—or industrial users—are able and willing to provide joint support. But there also are hundreds of substances that are, or could be, used in very limited volume to produce or process food such as narrow-spectrum pesticides, drugs, and flavor. Rarely can they justify substantial testing programs. To continue to add to the evidence of safety for those limited-volume chemicals society wishes to retain in use, or to develop useful new ones, we must find alternative approaches. These may include:

● Distributing the costs of safety evaluation as widely as possible among manufacturers, users, and government. Such an arrangement at least tends to spread the burden among a greater number of potential beneficiaries, although it may require some revision of antitrust laws and policies.

- Screening procedures (including short, inexpensive in vitro and in vivo tests) and assessments that help predict the relationship between structure or activity and the probable risk from exposure. Both approaches are coming into use as a means of obtaining preliminary indications of potential toxicity or risk of specific chemicals. By quickly providing important information at relatively little cost, they can direct later efforts toward areas of greater risk.

- Proposals for estimating and establishing "acceptable risk," which attempt to take into consideration such factors as social value and economic impact *(8–11)*, plus the consumer education efforts of the National Academy of Sciences *(12–14)*, the Institute of Food Technologists *(15)*, other organizations, and individuals *(16)*. These undertakings may result in evaluations better attuned to the actual needs and preferences of consumers who, once informed, can make some of their own judgements. (See the accompanying box on making decisions on food safety.)

The problem of inevitable incompleteness of any evaluation is equally difficult to manage. Hence, public discontent—and often resigned skepticism—are familiar reactions when newly discovered or assumed hazards are attributed to substances that have been used for years with apparent safety. Some of this new information comes simply from advances in toxicology and analytical chemistry. Testing today is more rigorous than ever before, employing larger numbers of animals, better controls, more thorough clinical and pathological examinations, and more sophisticated statistics. Sometimes these improved techniques turn up significant new evidence of potential hazards, although it is now widely accepted that every substance is hazardous if the intake level is high enough. Vitamins and amino acids, for example, are hazardous in very large quantities, and even drinking too much water can be fatal. As the search continues for more subtle and infrequent effects, occasional disconcerting findings are to be expected.

Since it is not possible to perform every conceivable test on every substance, particularly as the size, complexity, and number of tests continue to increase, at some point a use or no-use decision will have to be made on the available evidence because the low probability of obtaining additional information of value is unlikely to justify the additional effort needed to obtain it. Past judgments of this nature have usually been right, as evidenced by the large number of substances that have withstood repeated examination. In the early 1970s, the Food and Drug Administration (FDA) undertook a comprehensive review of the more than 600 additives in use before the 1958 Amendment. Most of these additives had not been explicitly pretested for safety, but were "generally regarded as safe"

Making Decisions on Food Safety

The Food Safety Council was formed in 1976 to develop new criteria and procedures for assessing the safety of foods.   Its board of trustees consists of an equal number of representatives from the food industry and from the public sector.   The Council's Social and Economic Committee has been seeking to develop a systematic decision-making structure that will provide informed, rational, and effective decisions regarding the food supply *(11)*.   In these efforts, the Council is trying to provide a framework for answering questions in several key areas:

Scientific Evidence
    Is there a demonstrated hazard?
    What is the nature of the hazard?
    What is the probability of exposure?
    Are there functional alternatives?

Social Value
    Will this maintain or improve the quality of life?
    Are there segmented population dimensions to this value?
    Is it culturally sound?

Economic Impact
    Will this increase the supply of food?
    What are the cost dimensions?
    Will there be international considerations?
    Will there be job dislocations?

The Decision Structure
    Who is to input into the process?
    What is the risk-to-benefit assessment?
    Who makes the decision?
    How is it promulgated?

Political Acceptability
    Are appropriate sectors of society represented?
    Can it be administered?
    Is it affordable?

(GRAS).   Of the almost 300 reviewed so far, about 80% have been reaffirmed as safe for current or foreseeable use without the need for new data *(17)*.   Most of the remainder were judged to pose no problems in their current uses.

The exceptions, however, are well publicized, disturbing, and often difficult to explain.   For example, the regulatory significance of the data on saccharin is controversial.   (For discussion of the saccharin question, see the section in Chapter 3 on Special Dietary Foods.)   Evaluating the safety of the food color Red No. 2 is hampered by the doubtful quality of some of the testing.   With both monosodium glutamate (MSG) and saccharin, there is a problem with test procedures not yet fully evaluated because there is little previous information to provide perspective and guidance.   Additionally, a new hazard may be perceived or an old one revived when a more sensitive analytical method finds previously undetectable traces—for example, diethylstilbestrol (DES) in beef liver, acrylonitrile in beverages in plastic bottles, and nitrosamines in cooked bacon.

Finally, there are a few instances in which ignorance or carelessness in the use of chemicals has clearly led to harm *(18)*.   Failure to foresee the consequences of use resulted in a minor but adverse reaction to an agent added to margarine to prevent spattering.   Two recent examples of carelessness are methyl mercury poisoning in humans when they accidentally consumed seed grain intended only for planting, and poisoning of cattle when feed was inadvertently mixed with polybrominated biphenyls, a toxic chemical used as a fire retardant *(18)*.

INCENTIVES FOR THE USE OF CHEMICALS

The costs of safety evaluation are a necessary part of the costs of development and are therefore related to the question of incentives—that is, the advantages or benefits to both producers and consumers associated with the use of the substance.   A number of advantages resulting from the use of specific chemicals in the food system have already been discussed.   But in a different sense, chemicals are used for the same reasons that many other substances and techniques are used—to satisfy individual wants and needs.   Foremost among those needs, of course, is health, including nutritional well-being and safety.   But, generally, foods are chosen from those available more for reasons of personal preference, which have primarily aesthetic, cultural, and social origins.

Incentives can be classified rather loosely into those related to economics, to health, and to aesthetic and psychological factors. The latter include cultural aspects, and all that might be called "quality of life." The different categories overlap and interact to a great extent. Economic benefits are, by definition, measured in dollars. Health benefits are not easily or even validly measured in dollars, and the value of a life-saving drug has no relationship to its cost. The only comprehensive, objective measure of "quality of life" benefits is what, in fact, people are willing and able to pay for them.

Thus the "value" of a chemical depends on the effects its use produces. These effects are both direct (first order), and indirect (second or lower order). An antioxidant that delays rancidity of a frying oil (direct effect) enables the oil to be used longer, with less waste (second order, economic benefit). The resulting food product keeps its flavor on the shelf for a longer period of time (a second order, quality-of-life benefit). As a consequence, the food processor will not need to deliver as frequently, nor pick up as many, over-age products (a third-order, economic benefit).

The price of the antioxidant will, of course, have to cover the costs of profitable manufacture and marketing, or no one will produce it. But what consumers will be willing to pay depends on how they evaluate the chain of benefits that follows from use of that antioxidant. Consumers do not evaluate these benefits "in a vacuum"; they do so in comparison with benefits that would flow from each available alternative. Some of these alternatives are in choice of chemicals or packaging, and others are procedural. A manufacturer may, for example, consider an alternative antioxidant, in which case he will need to compare the cost-effectiveness of the two (or more) choices in producing the benefits just discussed. Or, he may decide to abandon the use of any antioxidant, replace the oil more frequently, and raise the price. He may compensate for decreased shelf-life by more frequent deliveries (more expense) or by more protective packaging (more expense), or both. He may reformulate to use a more stable, more saturated oil (probably more expense, and possibly adverse health effects). Finally, if denied a critical ingredient by regulation or costs, the manufacturer may simply abandon the end product and take the economic loss. In all of this, he will attempt to follow, or preferably forecast correctly, how consumers respond to the values they perceive in the product.

Thus the real value is a net value that depends on the alternatives, among which the trade-offs are multiple, shifting, and complex. In summary, the real economic value of any chemical used in food production or processing should be measured, to the extent possible, by the total incremental costs of the next best alternative, which may be another chemical, a change in process, a shift in consumer use, or a complex mix of these.

EVALUATING BENEFITS AND RISKS UNDER THE LAW

The present legislative and regulatory system in the United States provides for the evaluation of risk, primarily by a regulatory agency such as FDA, and for evaluation of benefit, first by industrial users, but ultimately by the consumer in the marketplace.   However, a manufacturer wishing to use a chemical in food production or processing is necessarily also interested in risk and cannot afford to devote resources to the development of a substance unlikely to meet the regulatory agency's safety standards.

Federal   regulatory   agencies   have   traditionally   been   risk-oriented.   Risk, however, is narrowly construed.   Comparative risks are not considered, except informally and inexplicitly, since there is no statutory authority for their consideration.   If regulatory agencies consider benefits at all, they have tended to do so in a restricted set of cases or late in the evaluation process.   In the Food, Drug, and Cosmetic Act, Congress gave the Secretary of Health, Education, and Welfare only limited and highly structured authority—which he has delegated to the FDA—to consider benefits.   The so-called "Delaney Clause" (19) of the 1958 Amendment prohibits adding to foods any substance "found to induce cancer when ingested by man or animal" or found, in "appropriate" tests (other than those involving ingestion), to "induce cancer in man or animal."   The Delaney Clause excludes any consideration of benefit where cancer is a possible consequence of an *added* substance.   Section 402 of the Act, however, permits a "poisonous or deleterious substance"—including a carcinogen—if it is not added, and "if the quantity (of it) in food does not ordinarily render it injurious to health".   Thus very low levels of aflatoxin (one of the most potent carcinogens known for some animal species and thus suspect with respect to man) are permitted because the chemical is produced by a mold that may grow on certain foods.   In addition, the Delaney Clause does not apply to animal feed additives (19) and certain animal drugs (20) if they do not harm the animal and if they leave no residues in edible tissues when measured by prescribed methods.   Thus, where the benefit is the health of animals raised for food, FDA may consider that benefit in safety evaluation.

FDA may also consider benefit under Section 409, which says that where a safe level, generally referred to as a tolerance, must be set in order that a substance can be used safely, the tolerance shall not be higher than that required to achieve the intended effect.   The petitioner, consequently, must submit data establishing that such an intended effect (i.e., "benefit") does result.   Under Section 406 of the Act, the FDA can issue a regulation permitting the presence of an added "poisonous or deleterious substance" if it is "required in production" or "unavoidable in good manufacturing practice," or provided that the public health is still protected.   The terms "required" or "unavoidable" clearly imply benefit or else the foods produced could not be marketed.   The FDA is directed to consider the extent to which the substance is required or cannot be avoided.   These considerations of benefit in Section 409 and Section 406 affect the definition of adulterated food in Section 402.   Thus, while FDA may not generally judge the usefulness or benefit of food additives, there is an exception when safety is concerned.   It is a limited exception, however, since FDA may not judge whether the technical effect is desirable or not, but only whether it does occur and how much additive is needed to achieve it.

PUBLIC REACTION AND PUBLIC CONFIDENCE

The Food, Drug, and Cosmetic Act provided that as long as risk is not a significant factor, utility is best judged in the marketplace, where cost/benefit considerations are paramount.   But when society becomes aware of public health risks, especially with a substance long in use, special difficulties arise.   People are highly reluctant to accept any significant, definable, involuntary risks in their food supply, preferring simply to shift to an apparently no-risk alternative.   Unfortunately, there is always some risk—whether remote, indirect, or implied—in the use of every substance.   So, a shift to an apparently less hazardous alternative may only mean a shift to one whose risks have been evaluated less carefully.

Recent examples abound.   DDT was abandoned and replaced with demonstrably more toxic, though less persistent pesticides.   More methylene chloride is now used as an extraction solvent instead of trichloroethylene; more Red No. 40 is used instead of Red No. 2; and saccharin has replaced cyclamates as a nonnutritive sweetener.   It is now unclear whether any of these shifts has enhanced human health.   Because all these substances raise questions of carcinogenesis, the Delaney Clause applies.   The substances must be banned, and no consideration of benefit is possible.

The American public is uncomfortable with the comparison of economic benefits with vital risks ("How much will you spend to save a life?").   The public's discomfort is even greater with measures (other than personal choice) that seek to place a value on the quality of life, even though individuals daily make many decisions largely on this factor.   Health-related benefits usually involve, in fact, the reduction of some other risk—for example, milk fortified with vitamin D reduces the risk of rickets.

Aside from any consideration of economic or quality of life benefits, an action predicated solely on one health risk (a measurement of the risk of cancer) only means failure to compare other health risks. Society, in what seems at the moment the easy way out, takes a "piecemeal" approach instead of taking a more comprehensive view and recognizing that the hazards not yet known may be greater than those just recognized. It is only when other alternatives are exhausted, as in the case of saccharin, that society is forced to face up to the comparison of risks. It might be better to make this comparison earlier, preferably with a wider range of alternatives from which to choose so that the burden of use can be distributed. The Panel on Chemicals and Health of the President's Science Advisory Committee *(21)* suggested that government policy encourages a variety of chemical substances for every useful purpose. Such a variety has three advantages. It provides the opportunity to avoid overloading the body's capacity to deal with any one substance. It leaves regulatory agencies freer to move if new information indicates previously unsuspected hazards. And, finally, it serves the interest of the consumer by promoting competition and wider choice. However, not everyone considers it desirable to have a number of alternative substances for every useful purpose, especially with respect to food chemicals. Some governments—Great Britain for example—prefer to rely on a minimum number of supposedly thoroughly tested and evaluated substances.

A major problem in the safety evaluation process has its origin, as well as its possible solution, in chemistry. This problem stems from the fact that analytical skills to detect the presence of chemicals·are vastly ahead of capabilities to detect their toxicological effects, and these in turn are far ahead of ability to interpret certain of these effects with any real confidence in terms of human safety. Much of our current knowledge of the toxicity of chemicals comes from testing their effects in animals. Animal tests differ from real-life human exposure in several major respects: exposure level, species differences, genetic homogeneity of inbred laboratory animals in contrast to the genetic diversity of the human species, health, environment, and nutritional status. The interpretation of animal tests, therefore, is often problematical.

Greater confidence in appraising safety is not likely to come from animal feeding tests in which the end point is damaged tissues or dead animals. Such tests have value in demonstrating the existence of potential risk—or risk of a risk—but they provide no understanding of how it may occur, nor of its relevance to human exposure. Greater assurance and understanding are likely to come from studies in which the end point is an understanding of the disposition of a chemical and its metabolic products.

Chemistry, analytical chemistry in particular, plays a critical role in metabolic studies.   To be of any value, such studies in humans, as well as comparative metabolic studies in humans and other species that are animal models, must be conducted at intake levels relevant to the human diet, including some levels above normal intake to produce a margin of safety. Such studies usually look for very low levels of minor metabolic products at high dilution in body tissues and fluids.   To date, most of the advances in analytical chemistry have enhanced sensitivity.   Now simplicity and speed are also required.   Chemistry and biochemistry can provide a truer understanding of safety by determining what actually happens in the body—not merely the apparent safety based on human experience or gross measures from animal experiments.

CHEMISTRY AS A TOOL

Chemistry is a tool in the food system that involves not only substances but techniques and information.   It creates options, providing alternative substances and information that lead to wiser choices by government, individuals, or organizations.   Chemistry illuminates potential problems as well as potential solutions.   In the earlier days of gross adulteration and crude concepts of safety in our foods, new information led to awareness of problems and produced decisions that resulted in many obvious health and economic benefits.   As many of these major hazards have been eliminated, attention has shifted to more subtle ones.   At first, this new information generally led to new products and techniques, including means of anticipating or dealing with the undesirable effects of technology.   But more recently, new information has raised doubts about the safety of existing products and techniques.

Restraints are inherent in the use of chemicals.   Some begin at the design stage of a product—for example, the search for a nonpersistent pesticide that affects only specific species.   There may be limitations on the levels and conditions of use (as with some food additives), the physical forms in which the substance is applied (as with many pesticides), or conditions and purposes of use (as with drugs).

Use of a new chemical agent may be limited initially, allowing the consequences to be measured on a smaller scale before wide application. Almost invariably there are definitions and specifications to guard against unwanted impurities.   With food chemicals as well as with drugs, restraints may require monitoring of levels and effects. A number of these restraints will be reflected in the labeling of the substance or the foods that include it. An area of considerable controversy is whether users should be given all pertinent information and allowed to make their own choices, as is now done with cigarettes, or whether the restraints should be more positive, limiting choices or removing them entirely.

While chemistry has a key role to play in developing the information on which to make choices, in free societies these choices ultimately lie with the general public. They are made on the basis of what people are willing to pay for a particular product, service, or convenience, and what is perceived as an acceptable level of risk *(22)*. However, at the time that a government regulatory agency makes a decision that increases the cost or decreases the availability of a product, the ultimate consequences may not be publicly perceived. The controversy over whether saccharin can continue to be used in foods is a notable exception, precisely because some of the consequences of its elimination are so directly evident. Ultimately, to reduce apprehension and promote more effective and intelligent choices in the use of chemicals in the food system, more information will be needed to instill greater confidence, not only in technology but in all ramifications of its use by man.

REFERENCES, CHAPTER 5

1. Accum, Frederick *"A Treatise on Adulteration of Food and Culinary Poisons"*, 1820.
2. Federal Food, Drug, and Cosmetic Act of 1938; Public Law No. 717—75th Congress, Chapter 675, Third Session, S.5.
3. Federal Food, Drug, and Cosmetic Act, as amended; Pesticide Amendment of 1954, Section 408 (U.S.C. 21.346a).
4. Federal Food, Drug, and Cosmetic Act, as amended; Food Additives Amendment of 1958, Section 409 (U.S.C. 21.348).
5. Federal Food, Drug, and Cosmetic Act, as amended; Color Additives Amendments of 1960, Section 706 (U.S.C. 21.376).
6. Federal Food, Drug, and Cosmetic Act, as amended; Section 505 (U.S.C. 21.355) and Section 512 (U.S.C. 21.360b).
7. Toxic Substances Control Act, 40 C.F.R. Part 33 (U.S.C. 15.2601).
8. Clausi, A. "Coping With Issues of Food Safety," *Food Drug Cosmet. Law J.* **1978**, *33*, 372–375.
9. Turner, James S. "The Food Safety Council—One Small Step Toward a Better Way," *Food Drug Cosmet. Law J.* **1978**, *33*, 376–382.

10. Wodicka, Virgil O. "A System for Assessing the Safety of the Components of Food," *Food Drug Cosmet. Law J.* **1978**, *33*, 383–390.

11. Hopper, Paul F. "How Can We Make Better Decisions?" *Food Drug Cosmet. Law J.* **1978**, *33*, 391–395.

12. *"How Safe Is Safe? The Design of Policy on Drugs and Food Additives"*; National Academy of Science: Washington, DC, 1974.

13. *"Soil Fertility and the Nutritive Value of Crops"*; a statement of the Food and Nutrition Board, National Academy of Sciences, prepared by the Committee on Nutritional Misinformation, September, 1976.

14. "Food Safety Policy: Scientific and Societal Considerations"; part 2 of a 2-part study of the Committee for a Study of Saccharin and Food Safety Policy; National Academy of Sciences, Washington, DC, 1979.

15. Willey, Calvert L. "The IFT Public Information Program," *Food Technol. (Chicago)* **1975**, *29*(1), 16.

16. Lowrance, William W. "Of Acceptable Risk—Science and the Determination of Safety"; William Kaufman, Inc.: Los Altos, CA, 1976.

17. FASEB Select Committee on GRAS Substances. "Evaluation of Health Aspects of GRAS Food Ingredients: Lessons Learned and Questions Unanswered." *Fed. Proc., Fed. Am. Soc. Exp. Biol.* **1977**, *36*(11).

18. Johnson, Paul E. "Misuse in Foods of Useful Chemicals," *Nutri. Rev.* **1977**, *35*(9), 225.

19. Federal Food, Drug, and Cosmetic Act, As Amended; Section 409(c) (3) (A) (15.348).

20. Federal Food, Drug, and Cosmetic Act, As Amended; Section 512(d) (1) (H) (15.360b).

21. President's Science Advisiory Committee. "Chemicals and Health", Report of the Panel on Chemicals and Health; National Science Foundation, Washington, DC, 1973.

22. Wildavsky, Aaron. "No Risk Is the Highest Risk of All," *Am. Sci.* **1979**, January–February, 32–37.

Index

D

E

F

Book and jacket design by Carol Conway.

The book was composed by Carolina Academic Press,
Durham, North Carolina.

Printed and bound by Bookcrafters, Inc., Fredericksburg,
Virginia